Binary and Multiclass Classification

All questions and comments concerning this publication should be directed to publisher@weatherfordpress.com.

ISBN-13: 978-1-61580-016-2
ISBN-10: 1-61580-016-6

Thanks to my mother for guidance and support in pursuit of my dreams.

Table of Contents

1 Problems in Classification

1.1 Classification Problems

Classification problems arise when attempting to place a set of items into a predetermined set of categories. For example, we may put people in the categories of tall versus short depending on the height of each individual. When a person is taller than say 6', we classify them as tall. In this case, the classifier takes in as input the height of the person, and produces a binary output (0 or 1, true or false) by comparing the inputted height against the 6' standard.

These types of problems occur frequently in machine learning algorithms where a computer is programmed to examine characteristics of a set of inputs and place each input into one or more categories. For example, a computer program may examine a text file to determine the underlying language of the text. Here, the possible language choices are the categories, and the classifier is the computer algorithm that ingests texts and outputs the language detected.

Alternatively, we may have a medical test designed to identify the presence of a particular illness. The categories are simply the presence of absence of the illness, and the classifier is the medical test that takes as input a set of medical observations (symptoms, blood tests, etc.) and outputs a determination of the presence or absence of the illness.

Our goal is to understand the mathematics and statistics behind classification problems in order to evaluate the performance of classifiers. There are a wide variety of classification algorithms in use today. Naïve Bayesian classifiers, support vector machines, and neural networks are commonly used as classifiers. We are not concerned with the specifics of the underlying classification algorithms. Several excellent references are found in the bibliography that detail the specific implementation of these algorithms. Instead, we are concerned with understanding how to measure the performance of classifiers in order to better understand how reliable a given classifier is and how we can compare its performance against the performance of other classifiers.

We begin in this chapter by reviewing the binary classifier and multiclass classifier to create a foundation for later analysis. In the following chapters we examine the binary and multiclass classifiers in more detail. We also examine receiver operating characteristics as a means to evaluate the characteristics of a classifier and explore some statistics that arise in the measurement of classifiers.

1.2 Binary Classifiers

Binary classifiers classify inputs as either in the category or not in the category. For example, we can classify people as tall or not tall. In this case, a binary classifier takes a person's height as input and outputs either 'tall' or 'not tall'.

It is important to understand that a binary classifier is classifying as either in a category or not in the category. In many examples, we see two categories and the binary classifier placing inputs into one of the two categories. Carrying on with the above example, we may have tall versus short. A classifier may examine an input, and if it is determined that the person is tall, specify the category 'tall'. If the classifier determines the person is 'not tall', it may be assumed that the person belongs to the category 'short'. However, the binary classifier is really only specifying 'tall' or 'not tall'.

There is in fact only one category for a binary classifier. Inputs are classifies as either a member of the category or not. The specifics of the problem may determine that inputs that are not members of the category must be members of another category. But in fact, this process is itself a separate binary classifier.

This small but important difference often leads to misunderstandings with binary classifiers. In many practical cases, there are only two categories. For example, results of a simple coin toss are either 'heads' or 'tails'. A binary classifier will categorize as 'heads' or 'not heads', and 'not heads' must be equivalent to 'tails'. However, it should always be understood that the binary classifier is determining if the input belongs to the category or not. It is not choosing between two different categories.

Furthermore, viewing binary classifiers as placing inputs in a category or not allows for unusual input values. For example, in our simple coin toss, it is possible that a flip might result in the coin landing on edge. In this case, the binary classifier would result in 'not heads'. But this result is not the same as 'tails'. This is an unusual circumstance, but it is a possibility.

The ability of the classifier to respond to unusual circumstances can be a critical feature. In many cases of computerized classification, a binary classifier is applied to a very large set of inputs. Even though certain inputs may be highly unlikely (coin landing on edge), with sufficiently large data sets, very rare inputs may be present.

Assigning the negative results of a binary classifier (i.e. 'not tall') to another category should be approached with great care. Doing this is often a simplifying assumption which is very often true, but may be false in certain rare cases.

1.3 Multiclass Classifiers

Multiclass classifiers have more than one category. This is more than simply in or not in a single category. This means that there are multiple different categories, as well as a 'none of the above'.

1.3-a PIGEONHOLE CLASSIFIER

A pigeonhole classifier is a multiclass classifier where each input must be placed in exactly one of the categories (pigeonholes). The simplest form is a binary classifier with one category. However, generally a pigeonhole classifier is taken to mean a multiclass classifier.

The distinguishing feature of the pigeonhole classifier is that each input must be placed in exactly one category. Moreover, a NOTA pigeonhole classifier has a category for 'None of the Above' to handle cases where the classifier is unable to place an input into any other category. Alternatively, a NoNOTA pigeonhole classifier forces each input into exactly one of the categories, and there is no 'none of the above' category to choose.

1.3-b COMBINATION CLASSIFIER

Combination classifiers are multiclass classifiers that allow each input to be placed in multiple categories. Whereas the pigeonhole classifier places each input into exactly one category, the combination classifier places each input into one or more categories.

1.3-c FUZZY CLASSIFIER

Fuzzy classifiers place each input into one or more categories by degree. This does not necessarily mean that fuzzy logic is used as part of the classifier. Rather, instead of simply stating which categories an input belongs to (as in the combination classifier), the classifier computes a value (typically between 0 and 1) for each category. The result is a vector in the space of categories.

The output of fuzzy classifiers is often normalized in some manner. Commonly, the sum of the values of a given input over all categories is normalized to 1. Alternatively, if the values are treated as a vector, then we may normalize the magnitude of the vector to one rather than the sum of components.

1.3-d ESCAPE CLASSIFIER

Escape classifiers allow an input vector to not be placed in any category at all (one of your pigeons has flown the coop!). This may be used to modify any of the above classifiers. When used with a normalized fuzzy classifier, the input is simply ignored rather than attempting to normalize the result.

These are a few of the common multiclass classifiers. Other variants may arise in specific applications. However, whenever a classifier is designed, it is important to understand the specific properties of the categories. For example, an escape pigeonhole classifier may be distinct from a NOTA pigeonhole classifier. It may be the case where the escape is created by simply using the NOTA category as the escape. In this case, these two classifiers are equivalent. However, the NOTA category itself may have distinct meanings. NOTA may be a definitive statement that an input does not belong to any of the other categories. Alternatively, the NOTA category may be used to indicate that the classifier was unable to decide which one of a plurality of categories to classify the input. Thus, an escape pigeonhole classifier may have a NOTA category to categorize inputs that definitively do not belong in the other categories, and let escape inputs where the classifier was unable to choose from a plurality of categories.

Furthermore, multiclass classifiers can often be put in the form of a pigeonhole classifier. For example, suppose we use a NOTA combination classifier against three categories (A,B,C). An input may be placed in A, B, C, AB, AC, BC, ABC, or NOTA. Based on this we could construct a NOTA pigeonhole classifier on the seven categories: A, B, C, AB, AC, BC, and ABC.

1.4 Random Variables

We need to use some of the theory of random variables in order to analyze the performance of classifiers. In this section, we review the results we will need later.

1.4-a PROBABILITY

Suppose we have a random event X. We designate the probability of the occurrence of X as

$$P(X) \qquad\qquad 1.1$$

where P(X) is a value on the range [0,1]. For example, if we toss a fair coin, we may designate the probability of heads (H) versus tails (T) by writing

$$P(H) = .5$$
$$P(T) = .5 \qquad\qquad 1.2$$

Since the coin is fair, the probability of it turning up heads if the same as the probability for tails. As another example, the probability of drawing a jack (J) from a regular 52 card deck is

$$P(J) = \frac{4}{52} = \frac{1}{13} \qquad \text{1.3}$$

There are four jacks in the deck and 52 cards total.

Alternatively, we might ask what is the probability of not drawing a jack? There are 48 cards that are not a jack in 52 cards total. This gives,

$$P(\bar{J}) = \frac{48}{52} = \frac{12}{13} \qquad \text{1.4}$$

where the bar over the J indicates the negation (not a jack). But we should realize that we must draw either a jack (J) or not a jack ($\bar{J}$). In general, if we add the probability of an event to its negation, these must add to one:

$$P(X) + P(\bar{X}) = 1 \qquad \text{1.5}$$

We may also be interested in conditional probability. Suppose we draw a card from the deck and discard it. Then we draw a second card. What is the probability that the second card is a king given that the first card was a jack?

Since the first card was a jack and we discarded it, there are still four kings left in the deck, with 51 total cards remaining. Thus, the conditional probability is

$$P(K|J) = \frac{4}{51} \qquad \text{1.6}$$

The notation here tells us that we are computing the probability of drawing a king given that the first draw was a jack. Generalizing this, the probability of X occurring given Y is represented by

$$P(X|Y) \qquad \text{1.7}$$

Probabilities for two events may be combined in two ways. First, we need to be mindful that events are not always mutually exclusive. The events in flipping a coin are mutually exclusive. Each flip is either a heads or a tails, but never both. However, if we draw a card from a deck and consider both the card value and the suit, we may look at this either as 52 single events (one for each value-suit combination), or we may view this as two events (13 possible values in combination with 4 possible suits).

We ask the question, 'What is the probability of drawing a jack or a heart?' First, we examine this from the perspective of 52 single events. In this case, there are four jacks and 13 hearts with 52 total cards. However, one of the jacks is also a heart. This we have 16 different cards that match, so the probability is

$$P(J \cup H) = \frac{16}{52} = \frac{4}{13} \qquad 1.8$$

where P(X∪Y) means the probability of either X or Y occurring.

From the perspective of 13 values and 4 suits, we know that the probability of getting a particular value is $1/13$, and the probability of getting a particular suit is $1/4$. We want either of these events to occur. The probability of either of two events occurring is the sum of their probabilities, minus the probability of both occurring together.

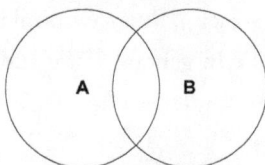

Figure 1: Events A and B where the events have some overlap.

Figure 2: The combination of both events A and B.

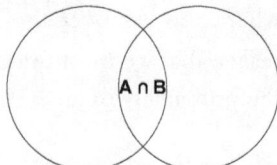

Figure 3: The intersection of events A and B.

We can visualize this from the Venn diagram in Figure 1. The figure shows two events, A and B, where the two events have some degree of overlap. If we want to compute the total area, we can add the area of A to the area of B. However, the region where A and B overlap is counted twice. We need to subtract the area of the intersecting region in order to get the correct value. Thus,

$$P(A \cup B) = P(A) + P(B) - P(A \cap B) \qquad 1.9$$

Where P(X∩Y) means the probability of both X and Y occurring.

Applying this to the question at hand, we know that the probability of drawing a jack is $1/13$, and the probability of drawing a heart is $1/4$. Also, the probability of drawing both a jack and a heart is $1/52$. Thus,

$$P(J \cup H) = \frac{1}{13} + \frac{1}{4} - \frac{1}{52} = \frac{4}{13} \qquad 1.10$$

We arrive at the same answer in both cases. In this case, since there are only 52 cards, it is easy to count up all the cards that are either jacks or hearts. However, in general, it may not be feasible to count in this way. The second method provides an alternative to approach problems where simple counting cannot be used.

In order to use our second method, we must be able to calculate the joint probability of two events occurring. Looking at Figure 3, we see that the probability of both A and B occurring is the probability of A alone occurring, multiplied by the probability of B occurring given A. In other words, we limit ourselves to just the circle A, then we count how many times B occurs within this circle. Of course, this works in both directions. We could also say that the joint probability is the probability of B times the probability of A given B. Mathematically,

$$P(A \cap B) = P(B)P(A|B) = P(A)P(B|A) \qquad \text{1.11}$$

Thus, in order to compute the joint probability, we can restrict ourselves to just one event, and then see how frequent the other event is within this group.

Concept	Expression
Probability of X	$P(X)$
Negation of X	$P(\bar{X})$
Probability of X given Y	$P(X|Y)$
Probability of X and Y	$P(X \cap Y) = P(Y)P(X|Y) = P(X)P(Y|X)$
Probability of X or Y	$P(X \cup Y) = P(X) + P(Y) - P(X \cap Y)$

Table 1: Basic probability concepts.

1.4-a(i) Bayes' Theorem

The expression for joint probability gives rise to Bayes' Theorem. If we examine the equation 1.11,

$$P(B)P(A|B) = P(A)P(B|A) \qquad \text{1.12}$$

or,

$$P(B) = \frac{P(A)P(B|A)}{P(A|B)} \qquad \text{1.13}$$

This allows us to write the probability of one event in terms of the probability of another event and the associated conditional probabilities.

1.4-b RANDOM VARIABLES

A random variable is a quantity whose value is determined through some probability. For example, we may have a random variable x where x=0 with a probability of 0.3 and x=1 with a probability of 0.7. In this case, the variable x can only have two values: 0 or 1. When we observe x, 30% of the time it has the value 0, and 70% of the time it has the value 1.

We can list all of the possible values for our random variable along with the probability that the value occurs. A graph for the previous case is shown in Figure 4. Similarly, Figure 5 plots the probability values for the random variable given in Table 2.

X	Probability	X	Probability
0	.1	3	.2
1	.04	4	.36
2	.3		

Table 2: Example discrete random variable tabulating the values of the variable (X) and the associated probabilities.

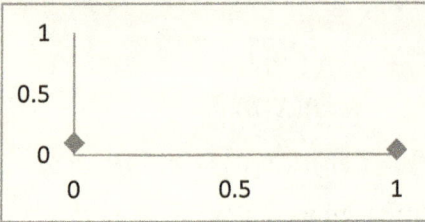

Figure 4: Graph of a random variable with value of 0 or 1 with probabilities .3 and .7 respectively.

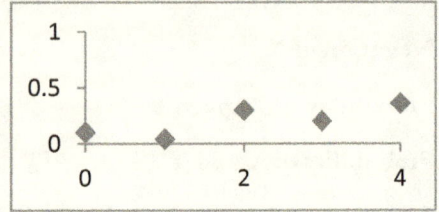

Figure 5: Graph of a random variable with value of 0 or 1 with probabilities .3 and .7 respectively.

These cases are discrete random variables because the random variable takes discrete values. In the first case, the variable takes the values 0 or 1, but never any number in between. Similarly, in the second case, the variable can take values 0, 1, 2, 3, or 4.

Alternatively, we can have continuous random variables. These are variables that take on continuous values (like the real number line). For example, the function

$$f(x) = \begin{cases} 2x & 0 < x < 1 \\ 0 & otherwise \end{cases} \qquad 1.14$$

may be used to describe a random variable. Here, $f(x)$ is the differential probability for the value x to occur. We need to use a differential probability because the random variable has an uncountable number of values. Because there are an uncountable number of values, the probability of drawing any particular single value is zero. However, the probability of obtaining a value over an uncountable range is not zero. The probability of obtaining a value for the random variable on the range (a, b) is given by the area under the curve of $f(x)$:

$$P(a < x < b) = \int_a^b f(x)dx \qquad \text{1.15}$$

We can use this concept to compute the probability of finding x somewhere between negative and positive infinity. Since this is all possible values of x, the probability of finding x on this range must be one:

$$P(-\infty < x < \infty) = \int_{-\infty}^{\infty} f(x)dx = 1 \qquad \text{1.16}$$

Thus, we must have

$$\int_{-\infty}^{\infty} f(x)dx = 1 \qquad \text{1.17}$$

At first this may appear to significantly restrict the possible functions $f(x)$ that may be used to describe a random variable. However, any function that has a finite integral can be used. Suppose we have a function $g(x)$ where

$$\int_{-\infty}^{\infty} g(x)dx = N \qquad \text{1.18}$$

We can turn this into a function describing a random variable by setting

$$f(x) = \frac{g(x)}{N} \qquad \text{1.19}$$

However, we must be mindful that any function describing a random variable can never be negative. Negative values indicate a negative probability for the variable, but a probability can never be negative.

We can create random variables that are a mix between discrete and continuous. We specify a function $f(x)$ that is nonzero and continuous in one region while having nonzero discrete values in other regions.

We can also create joint random variables where one variable is discrete while the other is continuous. We can create such a distribution by simply multiplying a discrete probability density function with a continuous probability density function to obtain a joint probability density where one of the variables is discrete and the other is continuous.

1.4-c DENSITY FUNCTION

The function in equation 1.15 is the probability density function for a continuous random variable. The probability density function is a common means to describe a random variable. The area under the curve of the probability density function between two points determines the probability of finding the value of the random variable over the range.

The probability density function can be used to compute the mean and variance of the distribution. The mean is computed as

$$\mu = \int_{-\infty}^{\infty} xf(x)dx \qquad 1.20$$

and the variance as

$$\sigma^2 = \int_{-\infty}^{\infty} (x-\mu)^2 f(x)dx \qquad 1.21$$

We can generalize this concept to higher powers of x. The nth moment of the distribution is

$$\mu_k = \int_{-\infty}^{\infty} (x-\mu)^k f(x)dx \qquad 1.22$$

The probability density function can be specified for a single variable random variable or for multiple random variables.

Probability Density Function for a Single Continuous Random Variable

Probability	$P(a < x < b) = \int_{a}^{b} f(x)dx$
Normalization	$\int_{-\infty}^{\infty} f(x)dx = 1$
Positive Definite	$f(x) \geq 0 \quad -\infty < x < \infty$
Mean	$\mu = \int_{-\infty}^{\infty} xf(x)dx$
Variance	$\sigma^2 = \int_{-\infty}^{\infty} (x-\mu)^2 f(x)dx$
n^{th} Moment	$\mu_k = \int_{-\infty}^{\infty} (x-\mu)^k f(x)dx$

Table 3: Probability density for a single continuous random variable.

Table 3 specifies the characteristics for the probability density function of a single random variable. Furthermore, Table 4 specifies the characteristics for multiple random variables.

In multivariate distributions, we can define cross moments between the different variables. For example, we define

$$k_{ij} = \int_{-\infty}^{\infty} \int_{-\infty}^{\infty} \cdots \int_{-\infty}^{\infty} (x_i - \mu_i)(x_j - \mu_j) f(x_1, x_2, \ldots, x_n) \, dx_1 dx_2 \ldots dx_n \qquad 1.23$$

as the covariance (second central moment). From this, we define the correlation coefficient

$$r_{ij} = \frac{k_{ij}}{\sqrt{\sigma_j^2 \sigma_j^2}} \qquad 1.24$$

The correlation coefficient is a measure of the linear relationship between the random variables x_i and x_j.

Probability Density Function for Multiple Continuous Random Variables

Probability	$P(a_1 < x_1 < b_1, a_2 < x_2 < b_2, \ldots, a_n < x_n < b_n)$ $= \int_{a_1}^{b_1} \int_{a_2}^{b_2} \cdots \int_{a_n}^{b_n} f(x_1, x_2, \ldots, x_n) \, dx_1 dx_2 \ldots dx_n$
Normalization	$\int_{a_1}^{b_1} \int_{a_2}^{b_2} \cdots \int_{a_n}^{b_n} f(x_1, x_2, \ldots, x_n) \, dx_1 dx_2 \ldots dx_n = 1$
Positive Definite	$f(x_1, x_2, \ldots, x_n) \geq 0 \quad -\infty < x_i < \infty$
Mean	$\mu_j = \int_{-\infty}^{\infty} \int_{-\infty}^{\infty} \cdots \int_{-\infty}^{\infty} x_j f(x_1, x_2, \ldots, x_n) \, dx_1 dx_2 \ldots dx_n$
Variance	$\sigma_j^2 = \int_{-\infty}^{\infty} \int_{-\infty}^{\infty} \cdots \int_{-\infty}^{\infty} (x_j - \mu(x_j))^2 f(x_1, x_2, \ldots, x_n) \, dx_1 dx_2 \ldots dx_n$
n^{th} Moment	$\mu_k^{(j)} = \int_{-\infty}^{\infty} \int_{-\infty}^{\infty} \cdots \int_{-\infty}^{\infty} (x_j - \mu_j)^k f(x_1, x_2, \ldots, x_n) \, dx_1 dx_2 \ldots dx_n$
Covariance	$k_{ij} = \int_{-\infty}^{\infty} \int_{-\infty}^{\infty} \cdots \int_{-\infty}^{\infty} (x_i - \mu_i)(x_j - \mu_j) f(x_1, x_2, \ldots, x_n) \, dx_1 dx_2 \ldots dx_n$
Correlation Coefficient	$r_{ij} = \frac{k_{ij}}{\sqrt{\sigma_j^2 \sigma_j^2}}$

Table 4: Probability density for multiple continuous random variables.

The probability density function determines the nature and behavior of the random variable. There are many important density functions used when working with random variables. However, many of these fundamentally originate from the discrete Binomial distribution.

The next sections examine the Binomial, Poisson, and Gaussian distributions. The Binomial and Poisson distributions are discrete distributions, while the Gaussian distribution is continuous. In fact, the Poisson distribution is an approximation to the Binomial distribution under one limit, while the Gaussian distribution is an approximation to the Binomial distribution under another limit.

1.4-c(i)   Discrete Distributions

The probability density function may be extended to discrete probabilities. When extending this concept, we replace the integrals in Table 3 and Table 4 with summations. This is

Probability Density Function for a Single Discrete Random Variable

Probability
$$P(a < x < b) = \sum_{x_i \in (a,b)} f(x_i)$$

Normalization
$$\sum_{i=0}^{N} f(x_i) = 1$$

Positive Definite
$$f(x_n) \geq 0 \quad 0 \leq n \leq N$$

Mean
$$\mu = \sum_{i=0}^{N} x_i f(x_i)$$

Variance
$$\sigma^2 = \sum_{i=0}^{N} (x_i - \mu)^2 f(x_i)$$

nth Moment
$$\mu_k = \sum_{i=0}^{N} (x_i - \mu)^k f(x_i)$$

Table 5: Probability density for a single discrete random variable with N events.

Probability Density Function for Multiple Discrete Random Variables

$$P(a_1 < x_1 < b_1, a_2 < x_2 < b_2, \ldots, a_n < x_n < b_n)$$

Probability $= \displaystyle\sum_{x_1 \in (a_1, b_1)} \sum_{x_2 \in (a_2, b_2)} \cdots \sum_{x_n \in (a_n, b_n)} f(x_1, x_2, \ldots, x_n)$

Normalization $\displaystyle\sum_{i_1=0}^{N_1} \sum_{i_2=0}^{N_2} \cdots \sum_{i_n=0}^{N_n} f(x_1, x_2, \ldots, x_n) = 1$

Positive Definite $f(x_1, x_2, \ldots, x_n) \geq 0 \quad -\infty < n < \infty$

Table 6: Probability density for multiple discrete random variables.

1.4-c(i).I Binomial Distribution

Suppose we have a recurring event (such as tossing a coin) where the probability of each occurrence of the event is p (p=.5 for tossing a fair coin). If we repeat the event multiple times, we obtain the Binomial distribution.

For example, suppose we have an event A that occurs with a probability p. On a single event, we get the distribution

$$f_1(n) = \begin{cases} n = 1 & p \\ n = 0 & 1 - p \end{cases} \qquad 1.25$$

where n is the number of occurrences of event A. With two events we have the distribution

$$f_2(n) = \begin{cases} n = 2 & p^2 \\ n = 1 & 2p(1 - p) \\ n = 0 & (1 - p)^2 \end{cases} \qquad 1.26$$

With two events, the probability that A occurs both times is p^2 (two heads in two tosses of a coin). The probability that A does not occur either time is $(1 - p)^2$. Finally, there are two ways we can get exactly one A: an A on the first event, and no A on the second, or no A on the first event and an A on the second. The probability of each of these is $p(1 - p)$, and the probability of either of these occurring is the sum of the two which is $2p(1 - p)$.

We can continue this process for additional event. As a final example, the distribution for three events is

$$f_3(n) = \begin{cases} n = 3 & p^3 \\ n = 2 & 3p^2(1 - p) \\ n = 1 & 3p(1 - p)^2 \\ n = 0 & (1 - p)^3 \end{cases} \qquad 1.27$$

In general, if there are k total events, the probability of obtaining n of event A is

$$P(n,k) = \binom{k}{n} p^n (1-p)^{k-n} \qquad 1.28$$

Using this, the probability distribution is

$$B_k(n) = \binom{k}{n} p^n (1-p)^{k-n} \quad 0 \le n \le k \qquad 1.29$$

Figure 6 shows a plot of the Binomial distribution with two events where the probability of A occurring is .3. With only two events, A occurs once with probability .3 and zero times with probability .7. Similarly, Figure 7 shows this situation with three events. Here, there is a probability of .49 that we get zero A's, .42 for 1 A, and .09 for two A's.

As the number of events increases, we see that the Binomial distribution has a peak followed by a tail. Figure 8 shows the Binomial distribution for ten events. At this point, we see the distribution increases to a peak, then tapers off.

Figure 9 shows the curve at thirty events. Here, the peak is even more distinguished. We see that most of the event counts are at nearly zero probability. Only the counts near the peak have much of a probability to occur. Also, note in these graphs that as the number of events increases, the height of the peak decreases.

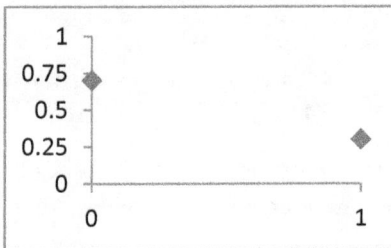

Figure 6: Graph of the Binomial distribution with two events where the probability of the event is .3.

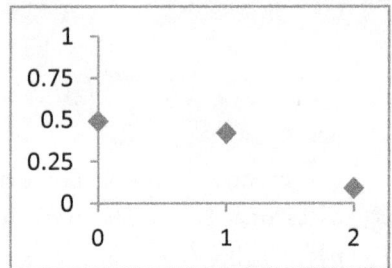

Figure 7: Graph of the Binomial distribution with three events where the probability of the event is .3.

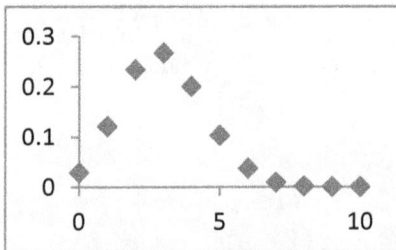

Figure 8: Graph of the Binomial distribution with ten events where the probability of the event is .3.

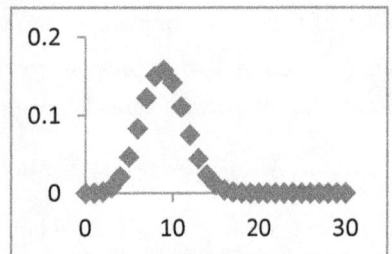

Figure 9: Graph of the Binomial distribution with thirty events where the probability is .3.

Notice that the width of the distribution near the peak decreases as the number of events increases. In addition, the point of the peak shifts to the right as the number of events increases. These properties are related to the mean and variance of the distribution.

Table 7 presents some basic properties of the Binomial distribution. There are two parameters governing the Binomial distribution. First, we have the number of trials N. Second, we have the probability p. The mean and variance of the binomial distribution are computed in terms of these parameters, as shown in the table.

Binomial Distribution Properties

Distribution	$B_N(n) = \binom{N}{n} p^n(1-p)^{N-n} \quad 0 \le n \le N$
Mean	Np
Variance	$Np(1-p)$

Table 7: Properties of the Binomial distribution.

1.4-c(i).II Poisson Distribution

The Poisson distribution is a discrete approximation to the Binomial distribution in the case where the number of trials is large, but the number of events Np remains small. Specifically, if the number of trials is large (N>50) and Np is small (Np < 5), the Binomial distribution is well approximated by the Poisson distribution:

$$p(x) = \frac{\lambda^x}{x!} e^{-\lambda} \qquad 1.30$$

where $\lambda = Np$. Table 8 provides some of the properties of the Poisson distribution.

Poisson Distribution Properties

Distribution	$p(x) = \dfrac{\lambda^x}{x!} e^{-\lambda} \quad N \ge 50, \lambda \le 5, \lambda = Np$
Mean	λ
Variance	λ

Table 8: Properties of the Poisson distribution.

1.4-c(ii)   Continuous Distributions

As the number of events in the Binomial distribution increases, we can approximate the density function using a continuous curve. The Gaussian distribution is a strong approximation when the number of events is large.

1.4-c(ii).I Gaussian Distribution

In the case where the Np is large (N > 50) and both Np and N(1-p) are large (Np > 5, N(1-p) > 5), the Binomial distribution is well approximated by the Gaussian distribution:

$$G(x) = \frac{1}{\sigma\sqrt{2\pi}} e^{-\frac{(x-\mu)^2}{2\sigma^2}}$$ 1.31

where $\mu = Np$ and $\sigma^2 = Np(1-p)$. Table 9 provides some of the properties of the Gaussian distribution.

Gaussian Distribution Properties

Distribution	$G(x) = \frac{1}{\sigma\sqrt{2\pi}} e^{-\frac{(x-\mu)^2}{2\sigma^2}}$ $N \geq 50, \mu \geq 5, \frac{\mu\sigma^2}{\mu-\sigma^2} \geq 5$ $\mu = Np, \sigma^2 = Np(1-p)$
Mean	Np
Variance	$Np(1-p)$

Table 9: Properties of the Gaussian distribution.

When the number of trials is large, the Binomial distribution is approximated by either the Poisson or Gaussian distributions. When Np is large, the Gaussian distribution makes a good approximation. Alternatively, when Np is small, the Poisson distribution is better suited.

1.4-c(ii).II   Beta Distribution

The Beta distribution is also related to the Binomial distribution. If we examine the Binomial distribution

$$B_N(k) = \binom{N}{k} p^k (1-p)^{N-k}$$ 1.32

This is a distribution telling us the conditional probability that k has a particular value given the value of p. Alternatively, we may express this as

$$P(k|p) = \binom{N}{k} p^k (1-p)^{N-k}$$ 1.33

Alternatively, we could fix k and think of this distribution as a function of p. In this case, we view p as a continuous variable on the range $[0,1]$. Since k is fixed, the combinatorial is just a constant number. We examine the distribution

$$P(p|k) = \bar{N}p^k(1-p)^{N-k} \qquad \text{1.34}$$

Where $\bar{N}$ is the normalization constant (we determine this by integrating over the range $[0,1]$ and setting the result to 1). Rather than using this particular form, we use the expression

$$P(p|k) = \bar{N}p^{\alpha-1}(1-p)^{\beta-1} \qquad \text{1.35}$$

in order to conform with the standard definition of the Beta distribution. Computing the normalization constant,

$$\int_0^1 \bar{N}p^{\alpha-1}(1-p)^{\beta-1}dp = 1 \qquad \text{1.36}$$

or

$$\bar{N} = \frac{1}{\int_0^1 p^{\alpha-1}(1-p)^{\beta-1}dp} \qquad \text{1.37}$$

The integral in the denominator is the beta function defined as

$$B(\alpha,\beta) = \int_0^1 p^{\alpha-1}(1-p)^{\beta-1}dp \qquad \text{1.38}$$

Using this, the probability density is

$$P(p|k) = \frac{p^{\alpha-1}(1-p)^{\beta-1}}{B(\alpha,\beta)} \qquad \text{1.39}$$

Beta Distribution Properties

Distribution	$Be(p) = \dfrac{p^{\alpha-1}(1-p)^{\beta-1}}{B(\alpha,\beta)} \quad 0 \le p \le 1$
Mean	$\dfrac{\alpha}{\alpha+\beta}$
Variance	$\dfrac{\alpha\beta}{(\alpha+\beta)^2(\alpha+\beta+1)}$

Table 10: Properties of the Beta distribution.

The Beta and Binomial distributions have similar forms. The Binomial distribution is the distribution of the number of times an even occurs (k) for a fixed probability for the event (p). The Beta distribution is the distribution of the probability for an event to occur (p) for a fixed number of events (k). The number of events must be an integer, so the Binomial distribution is a discrete distribution. Alternatively, since the probability for an event is on the range $(0,1)$, the Beta distribution is a continuous distribution.

With a Binomial distribution with N trials, k is on the range $[0, N]$. We can scale this to the range $[0,1]$ by examining the statistic $x = k/N$. We can find the mean of x from

$$\mu_x = \sum_{k=0}^{N} x B_N(k) \qquad\qquad 1.40$$

where $x = k/N$ and $B_N(k)$ is the Binomial distribution. Subsituting in the form of x,

$$\mu_x = \sum_{k=0}^{N} \frac{k}{N} B_N(k) \qquad\qquad 1.41$$

$$\mu_x = \frac{1}{N} \sum_{k=0}^{N} k B_N(k) \qquad\qquad 1.42$$

However, we know from Table 7 that the mean of the Binomial distribution is

$$\mu = Np \qquad\qquad 1.43$$

From the definition of the mean this is,

$$\mu = \sum_{k=0}^{N} k B_N(k) \qquad\qquad 1.44$$

Substituting these into 1.42,

$$\mu_x = \frac{\mu}{N} = p \qquad\qquad 1.45$$

Similarly, we can find the variance of x from

$$\sigma_x^2 = \sum_{k=0}^{N} (x - \mu_x)^2 B_N(k) \qquad\qquad 1.46$$

$$= \sum_{k=0}^{N} \left(\frac{k}{N} - \mu_x\right)^2 B_N(k) \qquad 1.47$$

$$= \frac{1}{N^2} \sum_{k=0}^{N} (k - N\mu_x)^2 B_N(k) \qquad 1.48$$

$$= \frac{1}{N^2} \sum_{k=0}^{N} (k - \mu)^2 B_N(k) \qquad 1.49$$

From Table 7, we know that the variance for the Binomial distribution is

$$\sigma^2 = Np(1 - p) \qquad 1.50$$

and from the definition,

$$\sigma^2 = \sum_{k=0}^{N} (k - \mu)^2 B_N(k) \qquad 1.51$$

Thus,

$$\sigma_x^2 = \frac{\sigma^2}{N^2} = \frac{p(1 - p)}{N} \qquad 1.52$$

The transformation $x = k/N$ scales the Binomial distribution to the range $[0,1]$. We can select a Beta distribution with the same mean and variance by setting

$$\frac{\alpha}{\alpha + \beta} = p$$
$$\frac{\alpha\beta}{(\alpha + \beta)^2(\alpha + \beta + 1)} = \frac{p(1 - p)}{N} \qquad 1.53$$

We can solve these equations for α and β. This is easier if we set

$$\alpha + \beta = s$$
$$\beta = s - \alpha \qquad 1.54$$

so that 1.53 becomes

$$\frac{\alpha}{s} = p$$
$$\frac{\alpha(s - \alpha)}{s^2(s + 1)} = \frac{p(1 - p)}{N} \qquad 1.55$$

From the first equation we see

$$\alpha = sp \qquad 1.56$$

Substituting into the second,

$$\frac{sp(s - sp)}{s^2(s + 1)} = \frac{p(1 - p)}{N} \qquad 1.57$$

or,

$$\frac{1}{(s + 1)} = \frac{1}{N} \qquad 1.58$$

so,

$$s = N - 1 \qquad 1.59$$

With s, we can find α and β as

$$\begin{aligned} \alpha &= (N - 1)p \\ \beta &= (N - 1)(1 - p) \end{aligned} \qquad 1.60$$

If we uses these values in the Beta distribution, the resulting probability density function will have the same mean and variance as a Binomial distribution with probability p on N trials.

1.4-d CUMULATIVE DISTRIBUTIONS

The cumulative distribution function is given by

$$P(y \leq x) = \int_{-\infty}^{x} f(y)dy \qquad 1.61$$

where $f(x)$ is the probability density function. The cumulative distribution is the probability that the random variable y is less than or equal to some value x. The cumulative distribution is useful when generating random variables.

The cumulative distribution must be increasing as x increases. In addition, the cumulative distribution must lie on the range $[0,1]$. Assume we have a cumulative distribution

$$C(x) = \int_{-\infty}^{x} f(y)dy \qquad 1.62$$

We can generate random variables that conform with the underlying probability density function $f(y)$ by generating a uniform random variable on the range $[0,1]$, then using the inverse of the cumulative distribution. Specifically, if u is a uniform random variable on the range $[0,1]$, then

$$C^{-1}(u) \qquad 1.63$$

is a random variable in accordance with the probability density function $f(y)$.

1.4-e FUNCTIONS OF A RANDOM VARIABLE

Suppose we have a random variable x generated in accordance with the probability density function $f(x)$. Now, suppose we have a function g such that

$$y = g(x)$$ 1.64

We want to compute the probability density function for y. Let $h(y)$ represent this probability density function. We can relate h to f and g by

$$h(y) = f(g^{-1}(y)) \left| \frac{dg^{-1}(y)}{dy} \right|$$ 1.65

This expression is valid so long as g is invertible. Alternatively, if g is not invertible, but has a countable number of roots, then we can write the probability density function for y in terms of the roots. Essentially, we sum the previous expression over all possible y's:

$$h(y) = \sum_i f\left(g_i^{-1}(y)\right) \left| \frac{dg_i^{-1}(y)}{dy} \right|$$ 1.66

1.4-f FUNCTIONS OF MULTIPLE RANDOM VARIABLES

We also examine functions of multiple random variables. Suppose we have two random variables x and y with joint probability density function

$$f(x, y)$$ 1.67

and functions

$$z = \bar{g}_z(x, y)$$
$$w = \bar{g}_w(x, y)$$ 1.68

with inverse functions

$$x = g_x(z, w)$$
$$y = g_y(z, w)$$ 1.69

We can find the joint probability density for z and w as

$$h(z, w) = f(g_x, g_y) |J(g_x, g_y)|^{-1}$$ 1.70

where J is the Jacobian of the transformation given by

$$J(x, y) = \begin{bmatrix} \dfrac{\partial \bar{g}_z}{\partial x} & \dfrac{\partial \bar{g}_z}{\partial y} \\ \dfrac{\partial \bar{g}_w}{\partial x} & \dfrac{\partial \bar{g}_w}{\partial y} \end{bmatrix} \qquad 1.71$$

1.4-g SUMS OF RANDOM VARIABLES

In later sections we examine the sum of random variables. In this section, we show that the distribution of the sum of two random variables is the convolution of the two distributions.

Suppose we have two distributions $f(x)$ and $g(y)$, governing random variables x and y. We want to find the distribution $h(z)$ for the sum of these variables $z = x + y$. If the variables are independent, their joint distribution is

$$D(x, y) = f(x)g(y) \qquad 1.72$$

Examine the variables

$$z = x + y$$
$$w = y \qquad 1.73$$

The inverse functions are

$$x = z - w$$
$$y = w \qquad 1.74$$

From 1.72, the joint distribution for z and w is

$$h(z, w) = f(z - w)g(w)|J|^{-1} \qquad 1.75$$

where

$$|J| = \begin{vmatrix} 1 & 1 \\ 0 & 1 \end{vmatrix} = 1 \qquad 1.76$$

Thus,

$$h(z, w) = f(z - w)g(w) \qquad 1.77$$

We find the distribution for z by integrating over w:

$$h(z) = \int_{-\infty}^{\infty} f(z - w)g(w)\, dw \qquad 1.78$$

This is the convolution of the original distributions.

As an example, examine the distribution of the sum of two random variables where both have a normal distribution with zero mean and unit variance. The distributions are

$$f(x) = \frac{1}{\sqrt{2\pi}} e^{-x^2/2} \qquad \text{1.79}$$

$$g(y) = \frac{1}{\sqrt{2\pi}} e^{-y^2/2} \qquad \text{1.80}$$

The distribution of the sum is

$$h(z) = \int_{-\infty}^{\infty} f(z-w)g(w)\,dw \qquad \text{1.81}$$

$$= \frac{1}{2\pi} \int_{-\infty}^{\infty} e^{-(z-w)^2/2} e^{-w^2/2}\,dw \qquad \text{1.82}$$

$$= \frac{e^{-z^2/2}}{2\pi} \int_{-\infty}^{\infty} e^{w(z-w)}\,dw \qquad \text{1.83}$$

$$= \frac{e^{-z^2/2}}{2\pi} \int_{-\infty}^{\infty} e^{-\left(w-\frac{z}{2}\right)^2 + \left(\frac{z}{2}\right)^2}\,dw \qquad \text{1.84}$$

$$= \frac{e^{-z^2}}{2\pi} \int_{-\infty}^{\infty} e^{-u^2}\,du \qquad \text{1.85}$$

$$= \frac{e^{-z^2}}{2\sqrt{\pi}} \qquad \text{1.86}$$

This is a normal distribution with zero mean and variance $\sigma^2 = \frac{1}{2}$.

1.4-h PROPAGATION OF ERRORS

Suppose we have measured a variable x and an associated error σ_x, and we have a function $f(x)$ and we need to compute the error associated with the function. We can use propagation of errors to find

$$\sigma_f^2 = \left(\frac{\partial f}{\partial x}\right)^2 \sigma_x^2 \qquad \text{1.87}$$

If we have a function of two measured variables,

$$\sigma_f^2 = \left(\frac{\partial f}{\partial x}\right)^2 \sigma_x^2 + \left(\frac{\partial f}{\partial y}\right)^2 \sigma_y^2 + 2\left(\frac{\partial f}{\partial x}\right)\left(\frac{\partial f}{\partial y}\right) k_{xy} \qquad \text{1.88}$$

where k_{xy} is the covariance between x and y as provided in Table 4.

2 Binary Classifiers

2.1 Simple Binary Classifier

The simple binary classifier (SBC) is the most basic of classifiers. SBCs have only two categories: A and Ā (not A). Simply put, the SBC determines that an input is either a member of the class (A) or not (Ā).

We need to be mindful that the classifier may be wrong. A given input may in fact be properly classified as A, but the SBC may return Ā. These errors can arise from a variety of sources: measurement error, recording error, transmission error, computer error, etc. We are not concerned with the specific nature of the errors, only in understanding how these errors affect the performance of the classifier.

Suppose we wish to test the performance of our SBC. We compile a set of inputs to present the SBC, where the correct classification is predetermined. We assume we have some oracle that is able to correctly classify this particular set of inputs. We present each of the inputs to the SBC and examine the resulting output. When we compare the classification from the oracle (Positive versus Negative for classification as A) with the output of the SBC, the SBC results will either be True (SBC matches the oracle) or False (SBC does not match the oracle).

The result is four possibilities:

1. True Positive – The SBC correctly classifies the input as A.
2. True Negative – The SBC correctly classifies the input as Ā.
3. False Positive – The SBC incorrectly classifies the input as A when in fact the input is Ā.
4. False Negative – The SBC incorrectly classifies the input as Ā when in fact the input is A.

We present each input to the SBC and count the number of times each of these results occurs. This may be presented as a confusion matrix (contingency table) as shown in Figure 10. In the figure, 'Predicted Outcome' is the result from the classifier while 'Actual Value' is the classification from the oracle.

For each input tested, exactly one of the four results from the confusion matrix must occur. Thus, the sum of the four values in the confusion matrix must equal the total number of inputs examined.

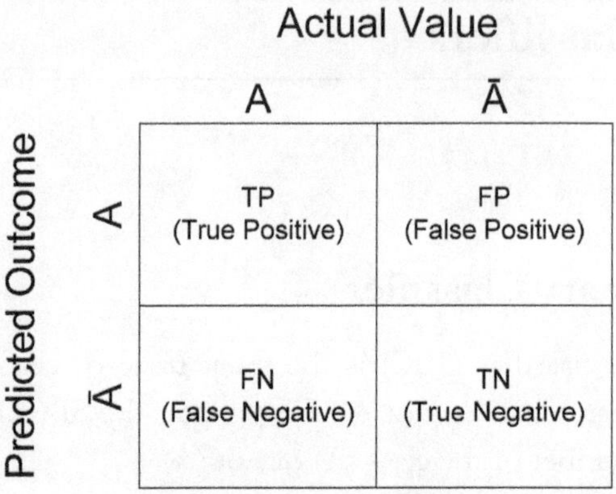

Actual Value

Figure 10: Confusion matrix for a simple binary classifier.

The confusion matrix is the fundamental measurement of the performance of a binary classifier. We present a classifier with a set of inputs that all have known results. In other words, we have some set of inputs where we already know whether each input is or is not in the category. Then we present this set of inputs to the classifier, and mark each input as correctly or incorrectly classified.

Based on this test, we can measure the classifier performance. Each input that the classifier correctly identifies as belonging to the category, we count as a true positive. Each input that the classifier correctly identifies as not belonging to the category we count as a true negative. Similarly, when the classifier incorrectly places an input in the category when the input truly is not in the category, we count these as false positives. Finally, when the classifier incorrectly determines that an input is not in the category when in fact we know that the input is in the category, we tally these as false negatives.

These results are then compiled as the confusion matrix as shown in Figure 10. These are the raw performance metrics of the binary classifier. We can create several different metrics based on these numbers. Each of these metrics provides different insight into a different aspect of the performance of the classifier. The next section will identify several such metrics and provide a mathematical expression that can be used to quantitatively measure a classifier's performance.

When performing a test of classifier performance, we should always keep in mind that the performance metrics we obtain are in relation to the input set provided. In creating an input set, we determine in advance how many inputs are in or not in the category. The classifier performance is measured in relation to this input set. An input set with a different mix of inputs may produce a different performance result.

2.2 Performance Metrics

Performance metrics for binary classifiers are important for comparing different classifiers. The confusion matrix may be used to compute several properties of a classifier. The following lists some of the most useful values resulting from the confusion matrix obtained after presenting N inputs to a SBC.

2.2-a TRUE POSITIVE (TP)

TP[1] is the number of times the SBC was correct when attempting to classify an input that truly belongs in the category. Here, the classifier determined that an input belongs in the category when in fact we know that the input really is in the category.

2.2-b FALSE POSITIVE (FP)

FP[2] is the number of times the SBC was not correct when attempting to classify an input that does not belong to the category. Specifically, the SBC determined an input is in A when in fact the input does not belong to A. This is the number of times the SBC misclassified an object as A.

2.2-c FALSE NEGATIVE (FN)

FN[3] is the number of times the SBC was not correct when attempting to classify an input that does belong to the category. Here, the SBC determined an input is not in A when in fact the input does belong to A.

2.2-d TRUE NEGATIVE (TN)

TN[4] is the number of times the SBC was correctly determined an input is not in the category.

---

[1] Also called hit

[2] Also called false alarm or Type I Error

[3] Also called miss or Type II Error

[4] Also called correct rejection

2.2-e ACTUAL POSITIVE (AP)

AP is the total number of inputs that in fact belong to the category. AP is found by taking the sum of the True Positives (we know these are actually true because the SBC classified them as true and the SBC was right) and the False Negatives (these are actually true because the SCB classified then as not belonging to the category and the SBC was wrong):

$$AP = TP + FN$$ 2.1

2.2-f ACTUAL NEGATIVE (AN)

AN is the total number of inputs that in fact do not belong to the category. Similar to the reasoning above, the Actual Negatives is the sum of the True Negatives and False Positives:

$$AN = FP + TN$$ 2.2

2.2-g CLASSIFIED POSITIVE (CP)

CP is the total number of inputs that the SBC classified in the category.

$$CP = TP + FP$$ 2.3

2.2-h CLASSIFIED NEGATIVE (CN)

CN is the total number of inputs that the SBC classifies as not in the category.

$$CN = FN + TN$$ 2.4

2.2-i TRUE POSITIVE RATE (TPR)

TPR[5] is the ratio of the number of times the SBC correctly determined an input should be in A to the total number of inputs that are in A. This is the ratio of the TP count to the total number of inputs that the SBC determined did not belong to the category (CP). Alternatively, this is also the conditional probably

---

[5] Also called sensitivity, hit rate, or recall

that the SBC correctly determines that an input belongs to the category given that the input is truly in the category.

$$TPR = P(C_A|T_A) = \frac{TP}{TP + FN} \qquad 2.5$$

where C_A is the event where the classifier determines that an input is in A, while T_A is the event that the input is actually in A.

2.2-j FALSE POSITIVE RATE (FPR)

FPR[6] is the ratio of the number of times the SBC failed to correctly identify an input should not be in A to the total number of inputs that are not in A. We can also interpret this as the probability that the SBC incorrectly classifies the input as in A given that the input is truly not in A:

$$FPR = P(C_A|T_{\bar{A}}) = \frac{FP}{FP + TN} \qquad 2.6$$

where C_A is the event where the classifier determines that an input is in A, while $T_{\bar{A}}$ is the event that the input is actually not in A.

2.2-k ACCURACY (ACC)

ACC is the ratio of the number of times the SBC correctly identified an input (either in A or not in A) to the total number of inputs tested. This is also the probability that the classifier correctly classifies the input.

$$ACC = \frac{TN + TP}{N} \qquad 2.7$$

2.2-1 TRUE NEGATIVE RATE (TNR)

TNR[7] is the ratio of the number of times the SBC correctly identified an input as not belonging to A to the total number of inputs that are not in A. This is the probability the SBC is correct when the SBC determines that an input is not in A.

$$TNR = P(C_{\bar{A}}|T_{\bar{A}}) = \frac{TN}{FP + TN} \qquad 2.8$$

---

[6] Also called fall-out
[7] Also called specificity

where $C_{\bar{A}}$ is the event where the classifier determines that an input is not in A, while $T_{\bar{A}}$ is the event that the input is actually not in A.

2.2-m FALSE NEGATIVE RATE (FNR)

TNR is the ratio of the number of times the SBC incorrectly identified an input as not belonging to A to the total number of inputs that are not in A. Alternatively, this is the probability that the SBC incorrectly classifies and not in A when the input is actually in A.

$$FNR = P(C_{\bar{A}}|T_A) = \frac{FN}{TP + FN} \qquad 2.9$$

where $C_{\bar{A}}$ is the event where the classifier determines that an input is not in A, while T_A is the event that the input is actually in A.

2.2-n POSITIVE PREDICTIVE VALUE (PPV)

PPV[8] is the proportion of SBC results that were correct when the SBC determined the input belongs to A. Alternatively, this is the probability that the input is actually in A given the SBC says that it is in A.

$$PPV = P(T_A|C_A) = \frac{TP}{TP + FP} \qquad 2.10$$

where C_A is the event where the classifier determines that an input is in A, while T_A is the event that the input is actually in A.

2.2-o NEGATIVE PREDICTIVE VALUE (NPV)

NPV is the proportion of SBC results that were correct when the SBC determined the input does not belong to A. This is also the probability that the input is not in A given that the SBC determines the same:

$$NPV = P(T_{\bar{A}}|C_{\bar{A}}) = \frac{TN}{TN + FN} \qquad 2.11$$

where $C_{\bar{A}}$ is the event where the classifier determines that an input is not in A, while $T_{\bar{A}}$ is the event that the input is actually not in A.

---

[8] Also called precision or positive predictive value

2.2-p FALSE DISCOVERY RATE (FDR)

FDR is the proportion of SBC results that were incorrect when the SBC determined the input belongs to A. This is the probability that the input is actually not in A given the SBC classifies the input as in A:

$$FDR = P(T_{\bar{A}}|C_A) = \frac{FP}{FP + TP}$$ 2.12

where C_A is the event where the classifier determines that an input is in A, while $T_{\bar{A}}$ is the event that the input is actually not in A.

2.2-q NON-DISCOVERY RATE (NDR)

NDR is the proportion of SBC results that were incorrect when the SBC determined the input does not belong to A. Alternatively, this is the probability that the input is in A given the SBC has classified it as not in A.

$$NDR = P(T_A|C_{\bar{A}}) = \frac{FN}{TN + FN}$$ 2.13

where $C_{\bar{A}}$ is the event where the classifier determines that an input is not in A, while T_A is the event that the input is actually in A.

2.2-r MATTHEWS CORRELATION COEFFICIENT (MCC)

MCC attempts to measure the overall quality of the SBC by accounting for the possibility of varying sizes of the categories (we will discuss this problem later).

$$MCC = \frac{(TPxTN - FPxFN)}{\sqrt{(TP + FP)(TP + FN)(TN + FP)(TN + FN)}}$$ 2.14

2.2-s F1 SCORE (F1)

F1 attempts to measure the overall quality of the SBC combining the PPV and TPR:

$$F_1 = 2\frac{PPVxTPR}{PPV + TPR}$$ 2.15

Each of these measure a different aspect of the performance of an SBC. In the next section, we will provide some examples and show what we can learn by examining the various measurements presented here. For reference, we present the performance metrics and their definitions in Table 12.

Metric Name	Abb.	Prob.	Definition
True Positive	TP	-	Count of the number of times the classifier is correct when the classifier determines an input belongs to the category.
False Positive	FP	-	Count of the number of times the classifier is not correct when the classifier determines an input belongs to the category.
False Negative	FN	-	Count of the number of times the classifier is not correct when the classifier determines an input does not belong to the category.
True Negative	TN	-	Count of the number of times the classifier is correct when the classifier determines an input does not belong to the category.
Actual Positive	AP	-	$AP = TP + FN$
Actual Negative	AN	-	$AN = FP + TN$
Classified Positive	CP	-	$CP = TP + FP$
Classified Negative	CN	-	$CN = FN + TN$
True Positive Rate	TPR	$P(C_A\|T_A)$	$TPR = \dfrac{TP}{TP + FN}$
False Positive Rate	FPR	$P(C_A\|T_{\bar{A}})$	$FPR = \dfrac{FP}{FP + TN}$
Accuracy	ACC	-	$ACC = \dfrac{TN + TP}{N}$
True Negative Rate	TNR	$P(C_{\bar{A}}\|T_{\bar{A}})$	$TNR = \dfrac{TN}{FP + TN}$
False Negative Rate	FNR	$P(C_{\bar{A}}\|T_A)$	$FNR = \dfrac{FN}{TP + FN}$
Positive Predictive Value	PPV	$P(T_A\|C_A)$	$PPV = \dfrac{TP}{TP + FP}$
Negative Predictive Value	NPV	$P(T_{\bar{A}}\|C_{\bar{A}})$	$NPV = \dfrac{TN}{TN + FN}$
False Discovery Rate	FDR	$P(T_{\bar{A}}\|C_A)$	$FDR = \dfrac{FP}{FP + TP}$
Non-Discovery Rate	NDR	$P(T_A\|C_{\bar{A}})$	$NDR = \dfrac{FN}{TN + FN}$
Matthews Correlation Coefficient	MCC	-	$\dfrac{(TPxTN - FPxFN)}{\sqrt{(TP + FP)(TP + FN)(TN + FP)(TN + FN)}}$
F1 Score	F1	-	$F_1 = 2\dfrac{PPVxTPR}{PPV + TPR}$

Table 11: Performance metrics definitions. In the probability column, P is the probability, C_A is the event that the classifier determines an input is in the category, $C_{\bar{A}}$ is the event that the classifier determines an input is not in the category, T_A is the event that an input is truly in the category, and $T_{\bar{A}}$ is the event that an input is truly not in the category.

2.3 SBC Examples

The quantities examined in the previous section measure different aspects of the performance of a SBC. To demonstrate this, we create some sample confusion matrices and compute these values. We examine these values to see what we may learn about the performance of a classifier based on these quantities.

2.3-a RANDOM CLASSIFIER

As a first example, suppose we have a SBC to which we present 100 inputs, and the result is 25 TP, 25 FP, 25 FN, and 25 TN. The characteristics of this classifier are presented in Table 12.

TP	25	FP	25
FN	25	TN	25
AP	50	AN	50
CP	50	CN	50
TPR	0.5	PPV	0.5
FPR	0.5	NPV	0.5
ACC	0.5	FDR	0.5
TNR	0.5	NDR	0.5
FNR	0.5	MCC	0
F1	0.5		

Table 12: SBC with 100 inputs with 25 TP, 25 FP, 25 FN, and 25 TN.

From the table, we presented the SBC with 100 inputs where 50 of them belonged to the class (AP) and 50 did not (AN). All of our measurements compute to a value of .5 with the exception of MCC which is zero. This is generally true when TP = FP = FN = TN.

The SBC classified 50 inputs as in the class (CP) and 50 as not in the class (CN). So far things look good. We know that there were 50 inputs that were in the category and 50 that were not. Furthermore, the classifier placed 50 inputs into the category and found 50 not. Is this classifier exhibiting perfect performance?

We look next at the TPR and FPR. TPR is .5, meaning that 50% of the inputs that are actually in the category were classified correctly by the SBC. Similarly, the FPR is .5 meaning that 50% of the inputs that are not in the category were correctly classified by the SBC.

The PPV and NPV show similar results. The PPV is 50% meaning that of the inputs classified by the SBC as in the category in fact belong to the category.

Moreover, the NPV at 50% means that of the inputs classified by the SBC as not in the category, half were correct.

The above numbers begin to paint the picture of the performance of our SBC. When an input belongs to A, the SBC correctly identifies this 50% of the time. Similarly, when an input does not belong to A, the SBC correctly identifies this 50% of the time. This is reinforced by the ACC which measures that the SBC is correct 50% of the time.

At this point our classifier appears to just make random guesses. How can we tell if this is what's going on? Let's examine how a random classifier would behave.

Assume we have a list of inputs of N inputs where the proportion ρ are actually in the category A while $1 - \rho$ are not in A. Suppose the classifier just randomly puts each input into a category with probability ε.

To determine how this classifier appears, compute the expected values of TP, FP, FN, and TN. First, we can compute AP and AN as:

$$AP = \rho N \qquad\qquad 2.16$$

$$AN = (1 - \rho)N \qquad\qquad 2.17$$

The classifier places each input randomly into the category with probability ρ. Using this we can compute CP and CN as:

$$CP = \varepsilon N \qquad\qquad 2.18$$

$$CN = (1 - \varepsilon)N \qquad\qquad 2.19$$

We can use this information to estimate the values of TP, FP, FN, TN for the ransom classifier. If we present a set of N inputs where ρN are actually in the category and the remainder not in the category, we can compute an estimate for the confusion matrix for a random classifier.

First, true positives arise when the input is actually in the category and the classifier correctly identifies the input as in the category. The total number of actual positives is ρN, and the classifier randomly assigns a proportion ε of these as belonging to the category. Hence,

$$TP = \varepsilon \rho N \qquad\qquad 2.20$$

Similarly, false negatives occur when the input is actually in the category but the classifier incorrectly determines that it is not. The random classifier assigns the proportion $(1- \varepsilon)$ of these as not belonging to the category. Thus, the estimate for the number of false negatives from the random classifier is

$$FN = (1 - \varepsilon)\rho N \qquad \text{2.21}$$

True negatives arise when the input is actually not in the category and the classifier correctly identifies this. There are $(1 - \rho)N$ inputs that actually do not belong to the category. Of these, the random classifier will randomly select a proportion $(1 - \varepsilon)$ as not in the category. Based on this we find the estimate of the number of true negatives as

$$TN = (1 - \varepsilon)(1 - \rho)N \qquad \text{2.22}$$

Finally, false positives arise when an input is actually not in the category, but the classifier incorrectly determines that the input should be in the category. Similar to the reasoning above we find

$$FP = \varepsilon(1 - \rho)N \qquad \text{2.23}$$

Using these results we can compute the metrics for the random classifier. Table 13 shows the various metrics for the random classifier.

We can invert these results to compare an arbitrary classifier to a random classifier. Set the variables

$$\varphi = \frac{AN}{AP} = \frac{1 - \rho}{\rho} \qquad \text{2.24}$$

$$\gamma = \frac{CN}{CP} = \frac{1 - \varepsilon}{\varepsilon} \qquad \text{2.25}$$

We invert these equations to find

$$\rho = \frac{1}{1 + \varphi} \qquad \text{2.26}$$

$$\varepsilon = \frac{1}{1 + \gamma} \qquad \text{2.27}$$

Comparing these equations to Table 12 we find

$$\varphi = \frac{AN}{AP} = 1 \qquad \text{2.28}$$

$$\gamma = \frac{CN}{CP} = 1 \qquad \text{2.29}$$

which means

$$\varepsilon = .5 \qquad \text{2.30}$$

$$\rho = .5 \qquad \text{2.31}$$

TP	$\varepsilon\rho N$	FP	$\varepsilon(1-\rho)N$
FN	$(1-\varepsilon)\rho N$	TN	$(1-\varepsilon)(1-\rho)N$
AP	ρN	AN	$(1-\rho)N$
CP	εN	CN	$(1-\varepsilon)N$
TPR	ε	PPV	ρ
FPR	ε	NPV	$1-\rho$
ACC	$1-\rho-\varepsilon+2\rho\varepsilon$	FDR	$1-\rho$
TNR	$1-\varepsilon$	NDR	ρ
FNR	$1-\varepsilon$	MCC	0
F1	$\dfrac{2\rho\varepsilon}{\rho+\varepsilon}$		

Table 13: Random SBC on N inputs with $\boldsymbol{a}$ proportion inputs in the class and $\boldsymbol{\rho}$ proportion classified as in members of the class.

If we substitute these values into the results from Table 13 and set N=100, we get the same results as shown in Table 12. Based on this, this SBC looks a lot like a random classifier.

Does this mean that the classifier from Table 12 is a random classifier? Not exactly. Even though all of the numbers work out exactly as we would expect a random classifier would, this does not mean the SBC from Table 12 is actually a random classifier. This could simply be a coincidence.

Does this at least mean that we cannot distinguish the SBC from Table 12 from a random classifier? This is true for this particular input set. However, if we were to construct a different input set, we might get very different results. We will discuss this more in Chapter 4.

In addition, in order to compare the results of two classifiers, we need to keep in mind that our performance measures are in fact experimental measurements. Each of these measurements has an associated error in the measurement. When comparing the results of two classifiers, we need to perform an error analysis to determine if the difference in the measured values is statistically significant.

We examine the error associated with the performance metrics in section 2.6. For now, it is important to understand that simply because one classifier has a higher performance metric than another classifier, this does not mean that the first classifier is better. The performance metrics, in isolation, are not enough to compare two classifiers. We need to examine the errors on the measurements, and perform statistical tests to quantify the significance of the differences between the two performance metrics.

2.3-b IDEAL CLASSIFIER

Next, suppose we have a SBC to which we present 100 inputs, and the result is 25 TP, 0 FP, 0 FN, and 75 TN. The characteristics of this classifier are presented in Table 12.

TP	25	FP	0
FN	0	TN	75
AP	25	AN	75
CP	25	CN	75
TPR	1	PPV	1
FPR	0	NPV	1
ACC	1	FDR	0
TNR	1	NDR	0
FNR	0	MCC	1
F1	1		

Table 14: SBC with 100 inputs with 25 TP, 0 FP, 0 FN, and 75 TN.

This classifier looks very promising from the start. We note that there are no false positives or false negatives. The TPR is 1 meaning that all of the inputs that are actually in the category are predicted to be in the category by the SBC. Furthermore, the PPV is also 1 which means that when the SBC predicts that an input is in the category, it is always actually in the category.

Similarly, the accuracy (ACC) is 1 as well. This means that every prediction that the classifier makes is correct. This classifier is never wrong. When the SBC predicts that an input is in the category, the input is actually in the category. When the SBC predicts that an input is not in the category, it is in fact not in the category.

This appears to be the ideal classifier. The values for the metrics in Table 14 are the 'perfect' values.

Does this mean that the classifier is in fact ideal? No. We need to be mindful that this is only the results of the classifier applied to this particular set of inputs. A different set of inputs may produce a different result.

Alternatively, the classifier may have simply been lucky. There is a finite probability, although small, that even a random classifier may have simply guessed right on every input.

We need to be careful about drawing broad conclusions about the performance of a classifier based on a single set of inputs. This classifier appears promising, but we should not infer that we have arrived at a perfect SBC.

2.3-c FALLACIOUS CLASSIFIER

Suppose we have a SBC where we present 100 inputs, and the result is 0 TP, 75 FP, 25 FN, and 0 TN. The characteristics of this classifier are presented in Table 12.

TP	0	FP	75
FN	25	TN	0
AP	25	AN	75
CP	75	CN	25
TPR	0	PPV	0
FPR	1	NPV	0
ACC	0	FDR	1
TNR	0	NDR	1
FNR	1	MCC	-0.57735
F1	∞		

Table 15: SBC with 100 inputs with 0 TP, 75 FP, 25 FN, and 0 TN.

This classifier looks very bad. We see right away that there are no true positives or true negatives. The TPR is 0 (none of the inputs that are actually in the category are predicted to be in the category by the SBC) and the PPV is also 0 (when the SBC predicts that an input is in the category, it is never right).

The accuracy (ACC) is 0 telling us that this classifier is always wrong. Every prediction that the classifier makes is incorrect. This classifier looks like the worst possible classifier we could imagine.

However, things may not be as bad as they seem. Imagine that we simply do the opposite of what the classifier predicts. In other words, when the classifier says that an input is in the category, we just switch the result to the opposite and say the input is not in the category. Similarly, when the classifier says the input is not in the category, we put it in. We negate the classifier.

What would the new classifier table look like? First, examine the effect on a false positive. If an input to a SBC results in a false positive, then the SBC predicted a positive when in fact the input was negative. If we negate the prediction of the SBC, then a positive prediction becomes a negative prediction. The actual value is still negative. We now have a negative prediction and an negative actual value, which is a true negative. Thus, by negating the classifier, false positives become true negatives.

Similarly, a false negative means that the original SBC predicted negative when the actual value is positive. By negating the prediction, we turn false negatives into true positives.

To complete the analysis, under negation, true positives become false negatives because the actual value is positive but we change the prediction from positive to negative. A negative prediction with a positive actual value creates a false negative. Similarly, true negatives become false positives under negation.

Figure 11 shows the effect of negation of the results of a classifier on the confusion matrix. We see that the movement of the elements of the confusion matrix is along the direction of the predicted outcome. There is no movement in the direction of the actual value. This is because we are changing the predicted outcome, but the actual value remains the same.

By negating the classifier, we transform true positives into false negatives (and vice versa) and transform false positives into true negatives (and vice versa). By applying negation to our original classifier, we transform from the classifier described in Table 15 to the classifier from Table 14 (the ideal classifier). We have transformed a classifier that is always wrong into a classifier that is always right!

Negation provides a very useful tool when examining classifiers. If we find a classifier that performs very poorly, we can simply negate the classifier to obtain a classifier that performs very well.

Based on this, we can create a duality concept for classifiers. By using classifier negation to form a dual, every classifier creates a pair of classifiers (the original classifier, and the negated classifier). If we always choose the more desirable of the two classifiers, we can obtain a classifier that demonstrates some level of performance. Based on this concept, we will show later that the worst possible classifier is the random classifier.

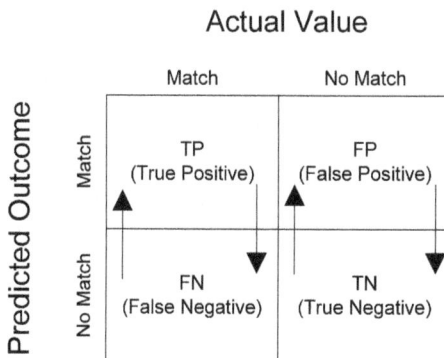

Figure 11: Confusion matrix for a negated classifier.

2.4 Constraint Analysis

In this section we examine how many values it takes to completely characterize a classifier for a given set of inputs. From section 2.1, we can construct a confusion matrix for a SBC when provided a set of inputs. Based on the confusion matrix, there are a variety of different values we can compute.

Each of the metrics from section 2.1 provides insight into a different aspect of the classifier. However, these metrics are all computed from the four values of the confusion matrix. Although these values are all useful in their own respect, they are not independent of each other.

When given a set of inputs, we present each input to our classifier. Based on these results, we compute the confusion matrix. This is the only data we have to characterize our classifier. Thus, we know that the most number of values we can use to characterize a classifier is four.

We can imagine plotting these values in a four dimensional space. Here, the four axes represent the values for TP, FP, FN, and TN. In this respect we say that our classifier is four-dimensional.

However, these four values have constraints, and these constraints reduce the dimension of the system. When we specify an input set with N inputs, there are a predetermined number of actual positives and actual negatives. Set the number of actual positives as ρN and the number of actual negatives as $(1 - \rho)N$.

When we test an input with our classifier, we must get a true positive, false positive, false negative, or true negative. We must get exactly one of these results. Thus, if we add up the values for these four elements of the confusion matrix, the result must be the total number of inputs:

$$TP + FP + FN + TN = N \qquad \text{2.32}$$

Next, examine the actual positives. We know there are ρN actual positives. If we present our classifier with an actual positive, and the classifier correctly identifies this as a positive, then the result is a true positive. Alternatively, if the classifier determines this input is a negative, then the result is a false negative. There is no other way to generate a true positive or false negative. Consequently, the sum of the true positives and false negatives must add to ρN.

$$TP + FN = \rho N \qquad \text{2.33}$$

We obtain a similar result for the actual negatives. There are $(1 - \rho)N$ actual negatives. When we present our classifier with an actual negative, we get either a true negative or a false positive. Therefore, the sum of the true negatives and false positives adds to $(1 - \rho)N$.

$$TN + FP = (1 - \rho)N \qquad 2.34$$

We have identified three constraints on our confusion matrix. However, the first constraint is simply the sum of the other two. Thus, we only have two linearly independent constraints on the confusion matrix.

This means that once we know either TP or FN, and either TN or FP, we can compute the other two values of the confusion matrix. With two constraints on four parameters, our four-dimensional system reduces to only two dimensions. Thus, we can completely characterize a classifier for a given input set with only two values.

It is important to understand that we are characterizing the classifier according to the input set used. A different input set may lead to different values. When we discuss characterizing a classifier, this should always be understood that we are characterizing with respect to a specific input set.

2.5 Negated Classifier

Earlier, we determined that a Fallacious Classifier can be made into an Ideal Classifier by simply negating the output of the classifier. Negation changes TP into FN and FP into TN and vice versa. This process also changes the performance metrics.

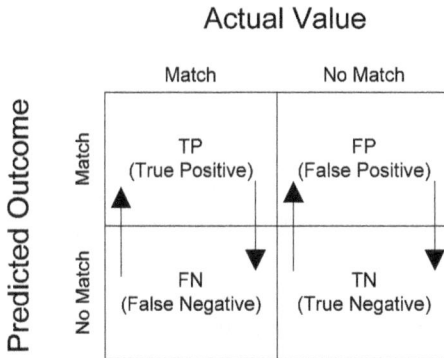

Figure 12: Confusion matrix for a negated classifier.

Figure 12 shows the how the confusion matrix for a binary classifier changes under negation. In addition to this, several of the performance metrics are interchanged as well.

Table 16 provides a list of how the metrics of a negated classifier relate to the metrics of the original classifier. For example, the PPV of a negated classifier has the same value as the NDR of the original classifier.

We determine this starting with the PPV for the original classifier:

$$PPV = P(T_A|C_A) = \frac{TP}{TP + FP} \qquad 2.35$$

since negation changes TP into FN and FP into TN. If we substitute this into the PPV we have

$$PPV_{Negated} = \frac{FN}{FN + TN} \qquad 2.36$$

But this is the same expression for the NDR of the original classifier. Thus, the PPV of the negated classifier has the same value as the NDR of the original classifier.

Original Metric	Negated Metric	Original Metric	Negated Metric
TP	FN	ACC	-
FP	TN	TNR	FPR
FN	TP	FNR	TPN
TN	FP	PPV	NDR
AP	AP	NPV	FDR
AN	AN	FDR	NPV
CP	CN	NDR	PPV
CN	CP	MCC	-
TPR	FNR	F1	-
FPR	TNR		

Table 16: Under negation, the performance metrics of a binary classifier are interchanged. The columns labeled 'Negated Metric' show how the negated classifier metrics relate to the original metrics. For example, the TPR for the negated classifier is the same as the FNR for the original classifier.

2.6 Distribution of Performance Measures

In the previous section, we found that the performance measures for a binary classifier are determined by only two values. We can choose one of TP / FN and one of FP / TN. Once we choose one of these from each group, the other two can be computed.

When we use the performance measures to compare classifiers, we need to know not only the value of the performance metric, but also the error associated with the metric. If one classifier has a metric of .8 and another value of .6, can we truly say that one performs better than the other? The error associated with these measurements is also required in order to make intelligent comparisons.

We use a particular input set to measure the performance of a classifier. Typically, we only do this test once. If we were to repeat the test again with a similar input set, we would expect that the precise values will be different, but the results will be 'similar'. What do we mean by 'similar'? If we computed a metric of .8 from one input set, and if we did this again with a similar input set, could we get .79? Or .7? Maybe even .4? Again, the error associated with the measurement provides insight to how these values might change if we performed a similar test with different inputs.

To determine the error, we need to know how the performance measurements are distributed. We use the theory of random variables to determine how we expect these values to change under different input sets.

For this section, we set

$$P = \rho N \qquad\qquad 2.37$$

$$F = (1 - \rho)N \qquad\qquad 2.38$$

where N is the total number of elements in the input set and ρ is the proportion of the input set that belongs to the category. Thus, P is the number of elements that should pass (belong to the category) while F is the number of elements that should fail. The values of P and F are independent of each other. This means that the values of TP / FN are independent of FP / TN.

Since we can measure the performance in terms of just two variables, we choose to use TP and FP as the values to characterize the classifier. We can compute FN and TN from

$$TP + FN = P \qquad\qquad 2.39$$

$$FP + TN = F \qquad\qquad 2.40$$

From these we can compute FN and TN from the values of P and F. If we set

$$\phi = \frac{TP}{P} \qquad\qquad 2.41$$

$$\chi = \frac{FP}{F} \qquad\qquad 2.42$$

we have

$$FN = (1 - \phi)P \qquad\qquad 2.43$$

$$TN = (1 - \chi)F \qquad\qquad 2.44$$

and

$$TP = \phi P \qquad\qquad 2.45$$

$$FP = \chi F \qquad\qquad 2.46$$

Both φ and χ are on the range [0,1]. Moreover, these are independent beta distributed variables. These variables are independent of each other because they result from independent binomial trials (P and F are independent of each other). The variable φ results from a binomial trial of the portion of the input set that belongs to the category (the P set), while χ results from a binomial trial of the portion of the input set that does not belong to the category (the F set). The intersection of the P and F sets is $\emptyset$ because a given input cannot be both in the category and not in the category.

We can use these expressions to formulate our performance measures in terms of φ and χ. For example, the true positive rate is

$$TPR = \frac{TP}{TP + FN} \qquad\qquad 2.47$$

Using 2.39,

$$TPR = \frac{TP}{P} \qquad\qquad 2.48$$

Substituting 2.45 into 2.48,

$$TPR = \varphi \qquad\qquad 2.49$$

Similarly, the false positive rate is

$$FPR = \frac{FP}{FP + TN} \qquad\qquad 2.50$$

$$FPR = \chi \qquad\qquad 2.51$$

The result for the PPV is more complicated. We have,

$$PPV = \frac{TP}{TP + FP} \qquad\qquad 2.52$$

$$= \frac{\varphi \rho N}{\varphi \rho N + \chi(1 - \rho)N} \qquad\qquad 2.53$$

$$= \frac{\varphi \rho}{\varphi \rho + \chi(1 - \rho)} \qquad \text{2.54}$$

$$= \frac{1}{1 + (\rho^{-1} - 1)\dfrac{\chi}{\varphi}} \qquad \text{2.55}$$

We can carry out similar procedures for the other performance measures. The results are shown in Table 17.

Metric	Formula	Metric	Formula
TP	$\rho \varphi N$	FP	$(1 - \rho)\chi N$
FN	$\rho(1 - \varphi)N$	TN	$(1 - \rho)(1 - \chi)N$
AP	ρN	AN	$(1 - \rho)N$
CP	$[\rho\varphi + (1 - \rho)\chi]N$	CN	$[\rho(1 - \varphi) + (1 - \rho)(1 - \chi)]N$
TPR	φ	PPV	$\dfrac{1}{1 + (\rho^{-1} - 1)\dfrac{\chi}{\varphi}}$
FPR	χ	NPV	$\dfrac{1}{1 + \dfrac{\rho}{1 - \rho}\dfrac{1 - \varphi}{1 - \chi}}$
ACC	$\rho\varphi + (1 - \rho)(1 - \chi)$	FDR	$\dfrac{1}{1 + \dfrac{\rho}{1 - \rho}\dfrac{\varphi}{\chi}}$
TNR	$1 - \chi$	NDR	$\dfrac{1}{1 + (\rho^{-1} - 1)\dfrac{1 - \chi}{1 - \varphi}}$
FNR	$1 - \varphi$	MCC	$(\varphi - \chi)\sqrt{\dfrac{\rho(1 - \rho)}{[\rho\varphi + (1 - \rho)\chi][\rho(1 - \varphi) + (1 - \rho)(1 - \chi)]}}$
F1	$\dfrac{2\varphi}{1 + \varphi + (\rho^{-1} - 1)\dfrac{\chi}{\varphi}}$		

Table 17: Performance measures in terms of φ and χ.

Each of the performance ratios can be written in terms of φ, χ, and α. The base performance measures are written in terms of φ, χ, α, and N. The performance ratios all have a factor of N in both the numerator and denominator.

Many of the performance measures from Table 17 can be written in the forms

TP / FP / TPR / FPR	$\lambda x + \bar{\lambda}y$	2.56

| FN / CP / CN / TN / ACC / TNR / FNR | $\lambda(1 - x) + \bar{\lambda}(1 - y)$ | 2.57 |

PPV / FDR	$$\dfrac{x}{x + \lambda y}$$	2.58
NPV / NDR	$$\dfrac{1 - x}{1 - x + \lambda(1 - y)}$$	2.59

where x and y are the performance measurements and λ and $\bar{\lambda}$ are constant coefficients. Table 18 lists the performance metrics in this form. All of the performance measures can be written as one of these four forms expect AP, AN, MCC, and F1. AP and AN do not depend on either φ or χ. F1 has a quadratic form, while MCC has a more complicated form. Both F1 and MCC attempt to create a single metric for measuring classifier performance. We are most interested in using two variables to measure classifier performance in order to get the full picture of the classifier behavior. As such, we will not examine the distributions of F1 and MCC.

Metric	Formula	Metric	Formula
TP	$\rho\varphi N$	FP	$(1 - \rho)\chi N$
FN	$\rho(1 - \varphi)N$	TN	$(1 - \rho)(1 - \chi)N$
AP	ρN	AN	$(1 - \rho)N$
CP	$[\rho\varphi + (1 - \rho)\chi]N$	CN	$[\rho(1 - \varphi) + (1 - \rho)(1 - \chi)]N$
TPR	φ	PPV	$\dfrac{\varphi}{\varphi + (\rho^{-1} - 1)\chi}$
FPR	χ	NPV	$\dfrac{1 - \chi}{1 - \chi + \dfrac{\rho}{1 - \rho}(1 - \varphi)}$
ACC	$\rho\varphi + (1 - \rho)(1 - \chi)$	FDR	$\dfrac{\chi}{\chi + \dfrac{\rho}{1 - \rho}\varphi}$
TNR	$1 - \chi$	NDR	$\dfrac{1 - \varphi}{1 - \varphi + (\rho^{-1} - 1)(1 - \chi)}$
FNR	$1 - \varphi$	MCC	$(\varphi - \chi)\sqrt{\dfrac{\rho(1 - \rho)}{[\rho\varphi + (1 - \rho)\chi][\rho(1 - \varphi) + (1 - \rho)(1 - \chi)]}}$
F1	$\dfrac{2\varphi^2}{\varphi + \varphi^2 + (\rho^{-1} - 1)\chi}$		

Table 18: Performance measures in terms of φ and χ.

We discussed in 1.4-c(ii).II, the Binomial distribution governs the case where we know p and want to find the probability of k occurrences in N trials. Alternatively, the Beta distribution governs the case where we have k successful outcomes and want to examine the distribution of p.

In our situation, we present N items to a classifier and measure the number of occurrences of TP and FP (effectively k), and we want to understand the distribution of φ and χ (effectively p). In fact, we have two distinct cases. First, we measure TP (k) and want to find the distribution of φ (p). Second, we measure FP (k) and want to find the distribution of χ (p). Thus, we have two independent Beta distributions: one governing φ, and the other governing χ.

Based on this, we want to find the distribution of our performance metrics. Our plan of attack is this:

1. Identify the distribution for our underlying variables φ and χ.
2. Apply the results of sections 1.4-e and 1.4-f to compute the distributions of 2.56-2.59.

We have already completed part of step 1: φ and χ have Beta distributions, but we don't know how to compute the parameters α and β for the distribution. Once we know the full form of the distribution, we can use sections 1.4-e and 1.4-f to compute the distribution of the functions in 2.56-2.59. This tells us how the performance measurement is distributed.

With everything complete, given N, ρ, TP, and PF for one classifier, we can compute the performance measures and error bounds. If we have these results for two classifiers, we can also compute the difference between the performance measures, the error bound on the difference, and determine if the difference is statistically significant.

2.6-a DISTRIBUTION OF THE UNDERLYING VARIABLES

The variables φ and χ are random variables under the Beta distribution. Given an input set with N total inputs where ρN of these belong in the category, and TP / P = φ, the parameters α and β are (see 1.60)

$$\alpha = \varphi(\rho N - 1) \qquad 2.60$$

$$\beta = (1 - \varphi)(\rho N - 1) \qquad 2.61$$

Similarly for χ, we have $(1 - \rho)N$ not in the category and FP / F = χ:

$$\alpha = \chi\big((1 - \rho)N - 1\big) \qquad 2.62$$

$$\beta = (1 - \chi)\big((1 - \rho)N - 1\big) \qquad 2.63$$

Be mindful that the distributions for φ and χ are independent, so the α and β for the φ distribution is different than the α and β for the χ distribution.

2.6-b Distribution of the Performance Metrics

We need to compute the distribution for metrics of the forms in equations 2.56-2.59. To start, look at the distribution of

$$\bar{\varphi} = 1 - \varphi \tag{2.64}$$

We know that φ has a Beta distribution:

$$\varphi \cong \frac{x^{\alpha-1}(1-x)^{\beta-1}}{B(\alpha,\beta)} \tag{2.65}$$

We can use equation 1.66 to compute the distribution of $\bar{\varphi}$:

$$h(y) = f\left(g^{-1}(y)\right)\left|\frac{dg^{-1}(y)}{dy}\right| \tag{2.66}$$

The original distribution $f(x)$ is

$$f(x) = \frac{x^{\alpha-1}(1-x)^{\beta-1}}{B(\alpha,\beta)} \tag{2.67}$$

The function g is the function from 2.64:

$$g(x) = 1 - x \tag{2.68}$$

We can find the inverse from

$$x = 1 - g^{-1}(x) \tag{2.69}$$

$$g^{-1}(x) = 1 - x$$

Substituting these into 2.66,

$$h(y) = f(1-y)|-1| \tag{2.71}$$

$$h(y) = \frac{(1-y)^{\alpha-1}(y)^{\beta-1}}{B(\alpha,\beta)} \tag{2.72}$$

But this is just another Beta distribution. If the original variable φ has a Beta distribution $B(\alpha,\beta)$, then $\bar{\varphi} = 1 - \varphi$ has the Beta distribution $B(\beta,\alpha)$. However, since $B(\beta,\alpha) = B(\alpha,\beta)$, the distribution of $\bar{\varphi}$ is the same as the distribution of φ.

This effectively reduces the four forms under consideration to just two:

$$\lambda x + \bar{\lambda}y \tag{2.73}$$

$$\frac{x}{x + \lambda y}$$
2.74

where both x and y are Beta distributed. Let x be a random variable with distribution function

$$x \cong \frac{x^{\alpha-1}(1-x)^{\beta-1}}{B(\alpha,\beta)}$$
2.75

and let y be a random variable with distribution function

$$y \cong \frac{y^{\bar{\alpha}-1}(1-y)^{\bar{\beta}-1}}{B(\bar{\alpha},\bar{\beta})}$$
2.76

Examining Table 18, there are several performance metrics that depend only on a single variable. We begin by examining the distribution of

$$z = \lambda x$$
2.77

From 1.66,

$$h(z) = f\left(g^{-1}(z)\right)\left|\frac{dg^{-1}(z)}{dz}\right|$$
2.78

$$= \lambda^{-1} f\left(\frac{z}{\lambda}\right)$$
2.79

$$= \lambda^{-1} \frac{\left(\frac{z}{\lambda}\right)^{\alpha-1}\left(1-\frac{z}{\lambda}\right)^{\beta-1}}{B(\alpha,\beta)}$$
2.80

This is the distribution governing a single variable. We see that when $\lambda = 1$, this distribution reduces to the original Beta distribution as expected.

We can use 1.70 to compute the distribution of the more general form

$$z = \lambda x + \bar{\lambda} y$$
2.81

Set

$$w = y$$
2.82

so

$$x = \lambda^{-1} z - \bar{\lambda}\lambda^{-1} w$$
2.83

$$y = w$$
2.84

The Jacobian is

$$J(x,y) = \begin{bmatrix} \dfrac{\partial z}{\partial x} & \dfrac{\partial z}{\partial y} \\ \dfrac{\partial w}{\partial x} & \dfrac{\partial w}{\partial y} \end{bmatrix} \qquad 2.85$$

$$= \begin{bmatrix} \lambda & \bar{\lambda} \\ 0 & 1 \end{bmatrix} \qquad 2.86$$

Thus,

$$|J(x,y)|^{-1} = \lambda^{-1} \qquad 2.87$$

The joint distribution of z and w is

$$h(z,w) = f(x,y)|J(x,y)|^{-1} \qquad 2.88$$

Since x and y are independent, their joint distribution is

$$f(x,y) = \frac{x^{\alpha-1}(1-x)^{\beta-1}y^{\bar{\alpha}-1}(1-y)^{\bar{\beta}-1}}{B(\alpha,\beta)B(\bar{\alpha},\bar{\beta})} \qquad 2.89$$

Substituting this into 2.88, and replacing the variables x and y using equations 2.83 and 2.84,

$$h(z,w)$$
$$= \frac{\left(\lambda^{-1}z - \bar{\lambda}\lambda^{-1}w\right)^{\alpha-1}\left(1 - \left(\lambda^{-1}z - \bar{\lambda}\lambda^{-1}w\right)\right)^{\beta-1}w^{\bar{\alpha}-1}(1-w)^{\bar{\beta}-1}}{B(\alpha,\beta)B(\bar{\alpha},\bar{\beta})}\lambda^{-1} \qquad 2.90$$

$$= \frac{\left(\lambda^{-1}z - \bar{\lambda}\lambda^{-1}w\right)^{\alpha-1}\left(1 - \lambda^{-1}z + \bar{\lambda}\lambda^{-1}w\right)^{\beta-1}w^{\bar{\alpha}-1}(1-w)^{\bar{\beta}-1}}{B(\alpha,\beta)B(\bar{\alpha},\bar{\beta})}\lambda^{-1} \qquad 2.91$$

$$= \frac{(\lambda^{-1}z)^{\alpha-1}(1-\lambda^{-1}z)^{\beta-1}(1-\bar{\lambda}z^{-1}w)^{\alpha-1}\left(1 - \dfrac{\bar{\lambda}\lambda^{-1}}{\lambda^{-1}z-1}w\right)^{\beta-1}w^{\bar{\alpha}-1}(1-w)^{\bar{\beta}-1}}{B(\alpha,\beta)B(\bar{\alpha},\bar{\beta})}\lambda^{-1} \qquad 2.92$$

Set

$$k = \frac{\bar{\lambda}\lambda^{-1}}{\lambda^{-1}z - 1} \qquad 2.93$$

$$l = \bar{\lambda}z^{-1} \qquad 2.94$$

We have,

$$h(z,w) = \frac{\lambda^{-1}l^{\alpha-1}(-k)^{\beta-1}(1-l)^{\alpha-1}(1-kw)^{\beta-1}w^{\bar{\alpha}-1}(1-w)^{\bar{\beta}-1}}{B(\alpha,\beta)B(\bar{\alpha},\bar{\beta})} \qquad 2.95$$

We want to find the distribution of z. We can find this from[9]

$$\bar{h}(z) = \int_0^1 h(z, w) dw \qquad 2.96$$

With this,

$$\bar{h}(z) = \frac{\lambda^{-1} l^{\alpha-1} (-k)^{\beta-1}}{B(\alpha, \beta) B(\bar{\alpha}, \bar{\beta})} \int_0^1 w^{\bar{\alpha}-1} (1-w)^{\bar{\beta}-1} (1-lw)^{\alpha-1} (1-kw)^{\beta-1} dw \qquad 2.97$$

We can evaluate the integral from[10]

$$\int_0^1 x^{\lambda-1} (1-w)^{\mu-1} (1-uw)^{-\rho} (1-vw)^{-\sigma} dx = B(\mu, \lambda) F_1(\lambda, \rho, \sigma, \lambda + \mu; u, v) \qquad 2.98$$

Thus,

$$\bar{h}(z) = \frac{\lambda^{-1} l^{\alpha-1} (-k)^{\beta-1} B(\bar{\beta}, \bar{\alpha})}{B(\alpha, \beta) B(\bar{\alpha}, \bar{\beta})} F_1(\bar{\alpha}, 1-\alpha, 1-\beta, \bar{\alpha} + \bar{\beta}; l, k) \qquad 2.99$$

or,

$$\bar{h}(z) = \frac{\bar{\lambda}^{\alpha+\beta-2}}{\lambda B(\alpha, \beta) z^{\alpha-1} (z-\lambda)^{\beta-1}} F_1\left(\bar{\alpha}, 1-\alpha, 1-\beta, \bar{\alpha} + \bar{\beta}; \bar{\lambda} z^{-1}, \frac{\bar{\lambda}}{z-\lambda}\right) \qquad 2.100$$

$$= \frac{\bar{\lambda}^{\alpha+\beta-2}}{\lambda B(\alpha, \beta) z^{\alpha-1} (z-\lambda)^{\beta-1}} F_1\left(\bar{\alpha}, 1-\alpha, 1-\beta, \bar{\alpha} + \bar{\beta}; \bar{\lambda} z^{-1}, \frac{\bar{\lambda}}{z-\lambda}\right) \qquad 2.101$$

where F_1 is the hypergeometric function.

This distribution is more complicated than the distribution in 2.80. We see that simply summing two Beta distributed variables results in a more complicated distribution than scaling a single variable.

Finally, we examine the distribution resulting from the form

$$z = \frac{x}{x + \lambda y} \qquad 2.102$$

Again we set

$$w = y \qquad 2.103$$

so that

$$x = \frac{\lambda wz}{1-z} \qquad 2.104$$

$$y = w \qquad 2.105$$

---

[9] The limits on the integral are determined from the possible values of w. Since w = y, and y is on the range [0,1], then w is also on the range [0,1].

[10] F_1 is the hypergeometric function on two variables. See Appendix A.1.

The Jacobian for this transformation is

$$J(x, y) = \begin{bmatrix} \dfrac{\partial z}{\partial x} & \dfrac{\partial z}{\partial y} \\[2ex] \dfrac{\partial w}{\partial x} & \dfrac{\partial w}{\partial y} \end{bmatrix} \qquad\qquad 2.106$$

Computing the partial derivatives:

$$\frac{\partial z}{\partial x} = \frac{x + \lambda y - x}{(x + \lambda y)^2} = \frac{\lambda y}{(x + \lambda y)^2} = \frac{(1 - z)^2}{\lambda w} \qquad\qquad 2.107$$

$$\frac{\partial z}{\partial y} = \frac{-\lambda}{(x + \lambda y)^2} = -\frac{(1 - z)^2}{\lambda w^2} \qquad\qquad 2.108$$

$$\frac{\partial w}{\partial x} = 0 \qquad\qquad 2.109$$

$$\frac{\partial w}{\partial y} = 1 \qquad\qquad 2.110$$

The determinant of the Jacobian is

$$|J(x, y)| = \begin{vmatrix} \dfrac{(1 - z)^2}{\lambda w} & -\dfrac{(1 - z)^2}{\lambda w^2} \\[2ex] 0 & 1 \end{vmatrix} \qquad\qquad 2.111$$

$$= \frac{(1 - z)^2}{\lambda w} \qquad\qquad 2.112$$

The joint distribution of z and w is

$$h(z, w) = f(x, y)|J(x, y)|^{-1} \qquad\qquad 2.113$$

$$= \frac{\left(\dfrac{\lambda w z}{1 - z}\right)^{\alpha - 1} \left(1 - \dfrac{\lambda w z}{1 - z}\right)^{\beta - 1} w^{\bar{\alpha} - 1}(1 - w)^{\bar{\beta} - 1}}{B(\alpha, \beta)B(\bar{\alpha}, \bar{\beta})} \frac{\lambda w}{(1 - z)^2} \qquad 2.114$$

$$= \frac{\lambda^\alpha z^{\alpha - 1} w^{\alpha + \bar{\alpha}}(1 - w)^{\bar{\beta} - 1}(1 - k w)^{\beta - 1}}{(1 - z)^{\alpha + 1}B(\alpha, \beta)B(\bar{\alpha}, \bar{\beta})} \qquad\qquad 2.115$$

where we have set

$$k = \frac{\lambda z}{1 - z} \qquad\qquad 2.116$$

We can find the distribution of z from

$$\bar{h}(z) = \int_0^1 h(z, w)\,dw \qquad\qquad 2.117$$

$$= \frac{\lambda^\alpha z^{\alpha-1}}{(1-z)^{\alpha+1}B(\alpha,\beta)B(\bar{\alpha},\bar{\beta})} \int_0^1 w^{\alpha+\bar{\alpha}}(1-w)^{\bar{\beta}-1}(1-kw)^{\beta-1}dw \qquad 2.118$$

We can evaluate the integral using

$$\int_0^1 z^{\beta-1}(1-z)^{\gamma-\beta-1}(1-kz)^{-\alpha}dz = B(\beta,\gamma-\beta)F(\alpha,\beta,\gamma;k) \qquad 2.119$$

Thus,

$$\bar{h}(z) = \frac{\lambda^\alpha z^{\alpha-1}}{(1-z)^{\alpha+1}B(\alpha,\beta)B(\bar{\alpha},\bar{\beta})}B(\alpha+\bar{\alpha}+1,\bar{\beta})F(1-\beta,\alpha+\bar{\alpha}+1,\alpha+\bar{\alpha}+\bar{\beta};k) \qquad 2.120$$

$$= \frac{\lambda^\alpha z^{\alpha-1}B(\alpha+\bar{\alpha}+1,\bar{\beta})}{(1-z)^{\alpha+1}B(\alpha,\beta)B(\bar{\alpha},\bar{\beta})}F\left(1-\beta,\alpha+\bar{\alpha}+1,\alpha+\bar{\alpha}+\bar{\beta};\frac{\lambda z}{1-z}\right) \qquad 2.121$$

Again, the distribution of the metric is expressed in terms of hypergeometric functions.

Thus far we have encountered three different distributions. Table 19 lists these distributions and the metrics that are governed by each distribution. Given we compute a performance metric, we can use the corresponding distribution to determine the range of values of the metric for a specific confidence level.

Distribution	Function	Metrics
$\mathfrak{D}_I(\alpha,\beta,\lambda;z)$	$\dfrac{z^{\alpha-1}(1-z)^{\beta-1}}{B(\alpha,\beta)}$	TP / FP / TPR / FPR / FN / TN / TNR / FNR
$\mathfrak{D}_{II}(\alpha,\beta,\bar{\alpha},\bar{\beta},\lambda,\bar{\lambda};z)$	$\dfrac{\bar{\lambda}^{\alpha+\beta-2}}{\lambda B(\alpha,\beta)z^{\alpha-1}(z-\lambda)^{\beta-1}}F_1\left(\bar{\alpha},1-\alpha,1-\beta,\bar{\alpha}+\bar{\beta};\bar{\lambda}z^{-1},\dfrac{\bar{\lambda}}{z-\lambda}\right)$	CP / CN / ACC
$\mathfrak{D}_{III}(\alpha,\beta,\bar{\alpha},\bar{\beta},\lambda,\bar{\lambda};z)$	$\dfrac{\lambda^\alpha z^{\alpha-1}B(\alpha+\bar{\alpha}+1,\bar{\beta})}{(1-z)^{\alpha+1}B(\alpha,\beta)B(\bar{\alpha},\bar{\beta})}F\left(1-\beta,\alpha+\bar{\alpha}+1,\alpha+\bar{\alpha}+\bar{\beta};\dfrac{\lambda z}{1-z}\right)$	PPV / FDR / NPV / NDR

Table 19: Distribution functions for binary classifier performance metrics.

Metric	Distribution $\mathfrak{D}_I$		
	α	β	λ
TP	$\varphi(\rho N - 1)$	$(1-\varphi)(\rho N - 1)$	ρN
FP	$\chi[(1-\rho)N - 1]$	$(1-\chi)[(1-\rho)N - 1]$	$(1-\rho)N$
TPR	$\varphi(\rho N - 1)$	$(1-\varphi)(\rho N - 1)$	1
FPR	$\chi[(1-\rho)N - 1]$	$(1-\chi)[(1-\rho)N - 1]$	1
FN	$\varphi(\rho N - 1)$	$(1-\varphi)(\rho N - 1)$	ρN
TN	$\chi[(1-\rho)N - 1]$	$(1-\chi)[(1-\rho)N - 1]$	$(1-\rho)N$
TNR	$\chi[(1-\rho)N - 1]$	$(1-\chi)[(1-\rho)N - 1]$	1
FNR	$\varphi(\rho N - 1)$	$(1-\varphi)(\rho N - 1)$	1

Table 20: Distribution functions for binary classifier performance metrics.

Metric	Distribution $\mathfrak{D}_{II}$					
	α	β	$\bar{\alpha}$	$\bar{\beta}$	λ	$\bar{\lambda}$
CP	$\varphi(\rho N - 1)$	$(1-\varphi)(\rho N - 1)$	$\chi[(1-\rho)N-1]$	$(1-\chi)[(1-\rho)N-1]$	ρN	$(1-\rho)N$
CN	$\varphi(\rho N - 1)$	$(1-\varphi)(\rho N - 1)$	$\chi[(1-\rho)N-1]$	$(1-\chi)[(1-\rho)N-1]$	ρN	$(1-\rho)N$
ACC	$\varphi(\rho N - 1)$	$(1-\varphi)(\rho N - 1)$	$\chi[(1-\rho)N-1]$	$(1-\chi)[(1-\rho)N-1]$	ρ	$(1-\rho)$

Table 21: Distribution functions for binary classifier performance metrics.

Metric	Distribution $\mathfrak{D}_{III}$				
	α	β	$\bar{\alpha}$	$\bar{\beta}$	λ
PPV	$\varphi(\rho N - 1)$	$(1-\varphi)(\rho N - 1)$	$\chi[(1-\rho)N-1]$	$(1-\chi)[(1-\rho)N-1]$	$(\rho^{-1}-1)$
FDR	$\chi[(1-\rho)N-1]$	$(1-\chi)[(1-\rho)N-1]$	$\varphi(\rho N - 1)$	$(1-\varphi)(\rho N - 1)$	$\dfrac{\rho}{1-\rho}$
NPV	$\chi[(1-\rho)N-1]$	$(1-\chi)[(1-\rho)N-1]$	$\varphi(\rho N - 1)$	$(1-\varphi)(\rho N - 1)$	$\dfrac{\rho}{1-\rho}$
NDR	$\varphi(\rho N - 1)$	$(1-\varphi)(\rho N - 1)$	$\chi[(1-\rho)N-1]$	$(1-\chi)[(1-\rho)N-1]$	$(\rho^{-1}-1)$

Table 22: Distribution functions for binary classifier performance metrics.

Finally, putting everything together, we can list the specific distribution for each of the performance metrics. Table 20-Table 22 specifies the parameters for each performance metric according to the distribution type.

2.7 Error Analysis

In the previous section we determined the distribution functions for each of the performance metrics of a binary classifier. In this section we examine statistical tests to determine when two classifiers perform differently. Even though the performance numbers for two classifiers may differ, they may be statistically the same.

As performance measures, we examine the values of φ and χ for the classifiers. Both φ and χ are treated as random variables under a Beta distribution. We use the variance of these variables as the measure of error (actually, the square of the error).

Let σ_φ^2 and σ_χ^2 represent the variance of these random variables. We can plot the values of φ and χ in two dimensions as shown in Figure 13. This plot presents the values for the metrics for this classifier, but it does not indicate the associated errors. We can place error bars around these values using σ_φ and σ_χ as the half-width of a box bounding the error (note: we use the square root of the variance as the half-width).

Since φ and χ are beta distributed, we can find the variance of these variables using the method of moments. Let N be the total number of inputs, and let ρN

be the proportion of actual positives. Then ρN is the number of trials associated with the parameter φ. Similarly, $(1-\rho)N$ is the number of trials associated with the parameter χ.

Comparing with equation 1.60 we find,

$$\alpha_\varphi = (\rho N - 1)\varphi$$
$$\beta_\varphi = (\rho N - 1)(1 - \varphi)$$

2.122

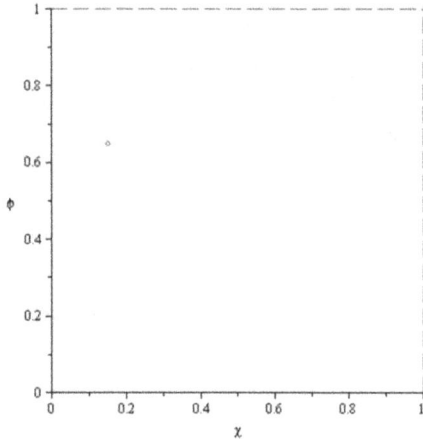

Figure 13: Plot of φ and χ values for a classifier where $\varphi = .65$ and $\chi = .15$.

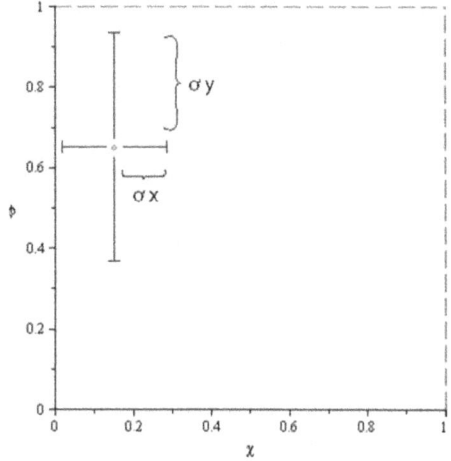

Figure 14: Plot of φ and χ values with error bars.

$$\alpha_\chi = [(1-\rho)N - 1]\chi$$
$$\beta_\chi = [(1-\rho)N - 1](1 - \chi)$$

2.123

The variance from these is

$$\sigma_\varphi^2 = \frac{\varphi(1-\varphi)}{\rho N}$$

2.124

$$\sigma_\chi^2 = \frac{\chi(1-\chi)}{(1-\rho)N}$$

2.125

We can use these expressions to compute the variance of our performance metrics. The variances are written in terms of the performance metrics, the number of inputs, and the proportion of inputs that are actually positive.

Figure 14 shows the same plot with error bars around both variables. This graphical aid helps indicate the magnitude of the error associated with each variable. However, the error bars only indicate the errors parallel to the axes. To fully understand how the error behaves, we need to compute the behavior of the error in the region surrounding the point.

2.7-a PROPAGATION OF ERRORS

One method to test the difference in the performance metrics of two points is to examine the Euclidean distance between the points relative to the error associated with the distance. The error in the distance may be computed using propagation of errors. We then examine the z-score (ratio of the distance to the square root of the variance) to determine if the distance is statistically significant.

The technique of propagation of errors assumes that the variables are all normally distributed. In many cases, the performance metrics are well approximated as a normal distribution. Using the z-score as a statistical test is only valid when the measurement is normally distributed. The assumption here is that the distribution of the distance value is a normally distributed variable.

Given a point (χ, φ) and associated variances σ_φ^2 and σ_χ^2, we can identify nearby points that are within a fixed error tolerance. Set τ be a specified error tolerance, and examine the uncertainty in the distance formula between (χ, φ) and another point $(\bar{\chi}, \bar{\varphi})$ with associated errors $\sigma_{\bar{\varphi}}^2$ and $\sigma_{\bar{\chi}}^2$. The distance between these points is

$$d = \sqrt{(\varphi - \bar{\varphi})^2 + (\chi - \bar{\chi})^2} \qquad\qquad 2.126$$

The error in this distance may be found through propagation of errors[11]

$$\sigma_d^2 = \left(\frac{\partial d}{\partial \varphi}\right)^2 \sigma_\varphi^2 + \left(\frac{\partial d}{\partial \chi}\right)^2 \sigma_\chi^2 + \left(\frac{\partial d}{\partial \bar{\varphi}}\right)^2 \sigma_{\bar{\varphi}}^2 + \left(\frac{\partial d}{\partial \bar{\chi}}\right)^2 \sigma_{\bar{\chi}}^2 \qquad 2.127$$

$$= \frac{(\varphi - \bar{\varphi})^2 \left(\sigma_\varphi^2 + \sigma_{\bar{\varphi}}^2\right) + (\chi - \bar{\chi})^2 \left(\sigma_\chi^2 + \sigma_{\bar{\chi}}^2\right)}{d^2} \qquad 2.128$$

Given two different classifiers, one with metrics (χ, φ) and errors $\left(\sigma_\chi^2, \sigma_\varphi^2\right)$, and the other with metrics $(\bar{\chi}, \bar{\varphi})$ and errors $\left(\sigma_{\bar{\chi}}^2, \sigma_{\bar{\varphi}}^2\right)$, the Euclidean difference in the values is

$$d = \sqrt{(\varphi - \bar{\varphi})^2 + (\chi - \bar{\chi})^2} \qquad\qquad 2.129$$

with uncertainty

$$\sigma_d = \frac{\sqrt{(\varphi - \bar{\varphi})^2 \left(\sigma_\varphi^2 + \sigma_{\bar{\varphi}}^2\right) + (\chi - \bar{\chi})^2 \left(\sigma_\chi^2 + \sigma_{\bar{\chi}}^2\right)}}{d} \qquad 2.130$$

---

[11] See § 1.4-g.

To test if these points are statistically different, we examine the z-score

$$z = \frac{d}{\sigma_d} \qquad 2.131$$

or,

$$z = \frac{(\varphi - \bar{\varphi})^2 + (\chi - \bar{\chi})^2}{\sqrt{(\varphi - \bar{\varphi})^2(\sigma_\varphi^2 + \sigma_{\bar{\varphi}}^2) + (\chi - \bar{\chi})^2(\sigma_\chi^2 + \sigma_{\bar{\chi}}^2)}} \qquad 2.132$$

The z-score is the measure of the statistic in relation to the standard error. This is a two-tailed statistics test, and when the z-score is greater than 1.96, then we are in the 95% confidence interval for statistical significance. A z-score of 1.96 is often used as a measurement of significance. However, other values may be used. In general, the relationship of the z-score to the significance level is

$$S = \sqrt{\frac{2}{\pi}} \int_0^z e^{-\frac{x^2}{2}} dx \qquad 2.133$$

Where S is the significance level and z is the z-score. Table 23 lists some commonly used values of the z-score and the associated significance levels.

		Significance			
z-score	.9	.95	.99	.995	.998
	1.65	1.96	2.58	2.81	3.08

Table 23: z-score v. significance for commonly used statistical tests.

We can use 2.132 to determine the locus of locus of nearby points that all have the same z-score. In this case, we treat $(\bar{\chi}, \bar{\varphi})$ as the measured point with errors $(\sigma_{\bar{\chi}}^2, \sigma_{\bar{\varphi}}^2)$. We want to find all points (χ, φ) that have the same z-score. The errors on the points (χ, φ) must be $(\sigma_\chi^2 = 0, \sigma_\varphi^2 = 0)$ because these points are exact (they are not measured points, just points on the coordinate system. The expression for the z-score becomes

$$z = \frac{(\varphi - \bar{\varphi})^2 + (\chi - \bar{\chi})^2}{\sqrt{(\varphi - \bar{\varphi})^2\sigma_{\bar{\varphi}}^2 + (\chi - \bar{\chi})^2\sigma_{\bar{\chi}}^2}} \qquad 2.134$$

Multiplying through by the denominator and squaring,

$$z^2[(\varphi - \bar{\varphi})^2\sigma_{\bar{\varphi}}^2 + (\chi - \bar{\chi})^2\sigma_{\bar{\chi}}^2] = [(\varphi - \bar{\varphi})^2 + (\chi - \bar{\chi})^2]^2 \qquad 2.135$$

If we make the coordinate change $x = \chi - \bar{\chi}, y = \varphi - \bar{\varphi}$,

$$z^2 \left[y^2 \sigma_{\hat{\varphi}}^2 + x^2 \sigma_{\hat{\chi}}^2 \right] = [y^2 + x^2]^2 \qquad 2.136$$

or,

$$(y^2 + x^2)^2 - z^2 \left(\sigma_{\hat{\varphi}}^2 y^2 + \sigma_{\hat{\chi}}^2 x^2 \right) = 0 \qquad 2.137$$

In general, this is a complicated curve. Figure 15 provides an example of the contours of this curve when x and y have unit error. The inner area is the region of points where z-score is less than 1.96. The points in this region are all within the 95% confidence bound. The outer area is the region of points where z-score is less than 2.58 (less than 99% confidence). The two regions together indicate how large the confidence bounds are and the additional area absorbed as we increase the confidence tolerance.

Figure 16 shows a similar situation where the error of y is twice the error of x. The curve is shaped similar to the intersection of two circles. Comparing with the previous graph, we see that the bounds of the confidence along the x-axis are the same. However, the bounds along the y-axis is twice the previous.

Figure 17 provides an example of how these error bounds might appear relative to a point. We see the same general shape as in Figure 16, however the bounds are centered on a particular performance point. Moreover, Figure 18 shows three points with error bounds. In this case, one of the points is separated from the others. However, two of the points demonstrate some degree of overlap. The 95% regions do not overlap, but the 99% regions do.

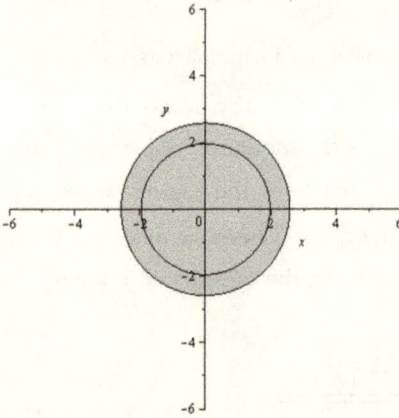

Figure 15: Plot constant z-scores where x and y have unit error. The inner region has z-score < 1.96, while the outer region has z-score < 2.58.

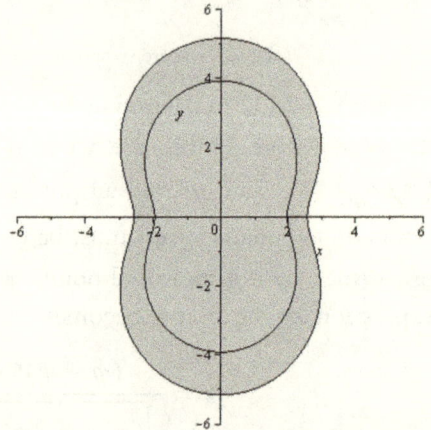

Figure 16: Plot constant z-scores where y has twice the error as x. The inner region has z-score < 1.96, while the outer region has z-score < 2.58.

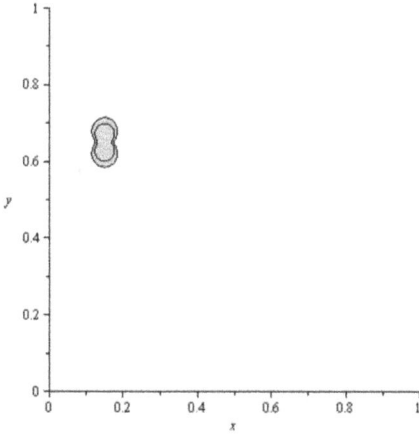

Figure 17: Example plot of classifier performance metrics with error bound.

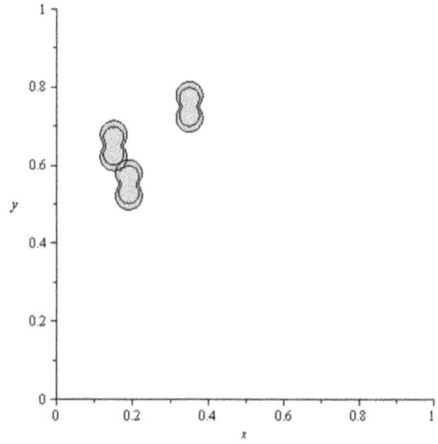

Figure 18: Example plot of multiple classifier points with error bounds.

Overlap of the 99% regions does not mean that these points are insignificantly different at the 99% level. In order to determine if the points are significantly different, we must use 2.134 and compute the z-score for the difference in the points in light of the relative errors. Similarly, if the regions do not overlap, this does not necessarily mean that the points are significantly different. Again, we must use 2.134 to definitively determine the significance.

The error bounds on the graph are meant to provide an indication of the regions of the z-scores. When error bounds highly overlap, this is a good indication that the points are not significantly different. When the error bounds are very far apart, this is a good indication that the points are significantly different. However, in any case, we should evaluate 2.134 to determine the significance.

2.7-b SUM OF Z-SCORES

The drawback of using the distance as a measure of significance is that if difference in one metric is much smaller than the other metric, then the second difference will dominate the distance. This is even worse when the variance associated with the first distance is much smaller than the second.

Another approach is to examine the sum of the z-scores for each individual metric. Here, each difference is measured relative to its own error, and only the ratio is considered. In this case, if a difference is small, but the variance is much smaller, the z-score is still large. By summing the z-scores we create a statistic that is able to relatively measure each contribution.

The z-score is

$$z = \frac{x}{\sigma_x} \tag{2.138}$$

where x is a measured value and σ_x^2 is the variance of the measurement. This test examines if the value x is significantly different from 0 (null hypothesis). The z-score is distributed as a Student-t distribution.

Suppose we measure x by sampling. Here, we take multiple independent measurements of x. Let x_i be the ith measurement. The mean and variance are approximated as

$$x = \frac{1}{N} \sum_{i=1}^{N} x_i \tag{2.139}$$

$$\sigma_x^2 = \frac{1}{N-1} \sum_{i=1}^{N} (x_i - x)^2 \tag{2.140}$$

If we use these values in equation 2.138, then the z-score is distributed as a Student-t distribution

$$f(t) = \frac{1}{\sqrt{v} B \left(\frac{1}{2}, \frac{v}{2} \right)} \left(1 + \frac{t^2}{v} \right)^{-\frac{v+1}{2}} \tag{2.141}$$

where v is the number of degrees of freedom. If we measure by sampling, $v = N - 1$. In most cases, we do not measure these by sampling. Instead, we measure the performance metric, then compute the variance by assuming the measurement is distributed according to a normal distribution. This is correct in the limit that we have an infinite number of measurements. Applying this here, we take the limit as $v \to \infty$. In this limit, the distribution becomes a normal distribution with unit variance:

$$\lim_{v \to \infty} f(t) = \frac{1}{\sqrt{2\pi}} e^{-\frac{t^2}{2}} \tag{2.142}$$

For this distribution, the probability that the absolute value of the z-score is greater than or equal to some value t is given by,

$$P(|z| \geq t) = \frac{2}{\sqrt{2\pi}} \int_{t}^{\infty} e^{-\frac{z^2}{2}} dz \tag{2.143}$$

$$= \frac{2}{\sqrt{\pi}} \int_{t/\sqrt{2}}^{\infty} e^{-z^2} dz \tag{2.144}$$

$$P(|z| \geq t) = erfc(t/\sqrt{2})$$ 2.145

where $erfc(t)$ is the complementary error function

$$erfc(t) = \frac{2}{\sqrt{\pi}} \int_t^\infty e^{-z^2} dz$$ 2.146

It is useful to note

$$\int_0^\infty e^{-z^2} dz = \frac{\sqrt{\pi}}{2}$$ 2.147

We use this relation when evaluating integrals later in this section.

Each z-score has an independent normal distribution. Examine the sum of two z-scores,

$$\bar{K} = x + y = \frac{\varphi - \bar{\varphi}}{\sqrt{\sigma_\varphi^2 + \sigma_{\bar{\varphi}}^2}} + \frac{\chi - \bar{\chi}}{\sqrt{\sigma_\chi^2 + \sigma_{\bar{\chi}}^2}}$$ 2.148

where both x and y are distributed according to 2.142. This sum treats both the x and y variables independently. This statistic is sensitive to both variables and is not dominated by large deviations in one difference relative to the other.

However, this statistic has a different problem. Both x and y can be either positive or negative. We can have a situation where x is large and negative, while y is large and positive, but the sum is zero.

Instead of this statistic, examine

$$K = |x| + |y|$$ 2.149

Now each term is positive so they can't cancel each other out.

Next, we analyze the behavior of this statistic. The variables x and y are independent of each other and normally distributed with unit variance. We can identify the distribution of $|x|$ from 2.142. Examine the statistic $z = |x|$ where x is distributed according to 2.142. We can write z as

$$z = \begin{cases} x & x \geq 0 \\ -x & x < 0 \end{cases}$$ 2.150

Solving this fox x,

$$x = \pm z$$ 2.151

From this, the distribution of z is

$$K_1(z) = \frac{1}{\sqrt{2\pi}} e^{-\frac{z^2}{2}} + \frac{1}{\sqrt{2\pi}} e^{-\frac{(-z)^2}{2}} \qquad 2.152$$

$$= \frac{2}{\sqrt{2\pi}} e^{-\frac{z^2}{2}} \qquad 2.153$$

This distribution is for $z \geq 0$. The full distribution is

$$K_1(z) = \begin{cases} \frac{2}{\sqrt{2\pi}} e^{-\frac{z^2}{2}} & z \geq 0 \\ 0 & z < 0 \end{cases} \qquad 2.154$$

With one variable, the probability for $K_1 \geq t$ is

$$P_1(K_1 \geq t) = \frac{2}{\sqrt{2\pi}} \int_t^\infty e^{-\frac{z^2}{2}} dz \qquad 2.155$$

$$= \frac{2}{\sqrt{\pi}} \int_{t/\sqrt{2}}^\infty e^{-z^2} dz \qquad 2.156$$

or

$$P_1(K_1 \geq t) = erfc(t/\sqrt{2}) \qquad 2.157$$

This matches the result from 2.146.

The subscript 1 above indicates we are examining the distribution from one such variable. With two independent variables, we use 1.78 to find the distribution of the sum:

$$K_2(z) = h(z) = \int_{-\infty}^\infty K_1(z - w) K_1(w) \, dw \qquad 2.158$$

The integrand vanishes whenever $w < 0$ or when $w > z$. Thus,

$$h(z) = \int_0^z K_1(z - w) K_1(w) \, dw \qquad 2.159$$

$$= \frac{2}{\pi} \int_0^z e^{-\frac{(z^2 - 2zw + 2w^2)}{2}} dw \qquad 2.160$$

$$= \frac{2}{\pi} e^{-\frac{z^2}{2}} \int_0^z e^{w(z-w)} dw \qquad 2.161$$

$$= \frac{2}{\pi} e^{-\frac{z^2}{4}} \int_0^z e^{-\left(w - \frac{z}{2}\right)^2} dw \qquad 2.162$$

$$= \frac{2}{\pi} e^{-\frac{z^2}{4}} \int_{-\frac{z}{2}}^{\frac{z}{2}} e^{-u^2} du \tag{2.163}$$

$$= \frac{4}{\pi} e^{-\frac{z^2}{4}} \int_{0}^{\frac{z}{2}} e^{-u^2} du \tag{2.164}$$

$$= \frac{2}{\sqrt{\pi}} e^{-\frac{z^2}{4}} erf(z/2) \tag{2.165}$$

The probability for $t \le K_2 < \infty$ is

$$P_2(K_2 \ge t) = \frac{2}{\sqrt{\pi}} \int_{t}^{\infty} e^{-\frac{z^2}{4}} erf(z/2) \, dz \tag{2.166}$$

$$= \frac{4}{\sqrt{\pi}} \int_{t/2}^{\infty} e^{-z^2} erf(z) \, dz \tag{2.167}$$

$$= 1 - erf^2(t/2) \tag{2.168}$$

2.7-c HOTELLING'S T-SQUARE

Instead of using the sum absolute value of the z-score, we could examine the sum of the squares of the z-scores. This approach is similar to Hotelling's T-Square. Hotelling's T-Square applies to a set of measurements of different dependent variables. The technique can be extended to sets with an infinite degree of freedom on multiple dependent variables.

In the previous section, we found that the z-score is distributed as a normal distribution with zero mean and unit variance

$$f(t) = \frac{1}{\sqrt{2\pi}} e^{-\frac{t^2}{2}} \tag{2.169}$$

The distribution of $z = t^2$ may be computed from 1.66:

$$H_1(z) = f(\sqrt{z}) \left| \frac{1}{2\sqrt{z}} \right| + f(-\sqrt{z}) \left| \frac{1}{2\sqrt{z}} \right| \tag{2.170}$$

$$H_1(z) = \frac{1}{\sqrt{2\pi}} z^{-\frac{1}{2}} e^{-\frac{z}{2}} \tag{2.171}$$

This is a χ^2 distribution on one degree of freedom. The probability that z exceeds a value t is

$$P_1(H_1 \ge t) = \frac{1}{\sqrt{2\pi}} \int_{t}^{\infty} z^{-\frac{1}{2}} e^{-\frac{z}{2}} dz \tag{2.172}$$

$$= \frac{1}{\sqrt{2\pi}} \gamma \left(\frac{1}{2}, \frac{x}{2}\right) \tag{2.173}$$

where γ is the lower incomplete gamma function.

The sum of two variables distributed according to 2.171 can be found from the convolution of the distribution:

$$H_2(z) = \int_{-\infty}^{\infty} H_1(z - w) H_1(w) \, dw \tag{2.174}$$

$$= \frac{1}{2\pi} \int_{-\infty}^{\infty} (z - w)^{-\frac{1}{2}} e^{-\frac{(z-w)}{2}} w^{-\frac{1}{2}} e^{-\frac{w}{2}} \, dw \tag{2.175}$$

$$= \frac{1}{2\pi} e^{-\frac{z}{2}} \int_{-\infty}^{\infty} (z - w)^{-\frac{1}{2}} w^{-\frac{1}{2}} \, dw \tag{2.176}$$

$$= \frac{1}{2\pi} z^{-1} e^{-\frac{z}{2}} \int_{-\infty}^{\infty} \left(1 - \frac{w}{z}\right)^{-\frac{1}{2}} \frac{w^{-\frac{1}{2}}}{z} \, dw \tag{2.177}$$

$$= \frac{1}{2\pi} e^{-\frac{z}{2}} \int_{-\infty}^{\infty} (1 - u)^{-\frac{1}{2}} u^{-\frac{1}{2}} \, du \tag{2.178}$$

$$= \frac{1}{2} e^{-\frac{z}{2}} \tag{2.179}$$

This is also a χ^2 distribution, but this is on two degrees of freedom. The probability that z exceeds a value t is

$$P_2(H_2 \geq t) = \frac{1}{2} \int_t^{\infty} e^{-\frac{z}{2}} dz \tag{2.180}$$

$$= \gamma \left(1, \frac{t}{2}\right) \tag{2.181}$$

From these results, we may examine the statistic

$$H = x + y = \frac{(\varphi - \bar{\varphi})^2}{\sigma_\varphi^2 + \sigma_{\bar{\varphi}}^2} + \frac{(\chi - \bar{\chi})^2}{\sigma_\chi^2 + \sigma_{\bar{\chi}}^2} \tag{2.182}$$

We compute this value for a particular binary classifier. The confidence level of difference is determined from P_2:

$$P_2(H_2 \geq H) = \gamma \left(1, \frac{H}{2}\right) \tag{2.183}$$

Here, we use the lower incomplete gamma function to determine the probability that we would find a value of H simply by random chance. A table of the regularized gamma function is provided in Appendix E.

2.8 Duals

If we think of measuring the classifier performance in terms of φ and χ rather than in terms of TP and FP, we see that many of our performance measures are related under the mapping

$$\varphi \to \chi \qquad \text{2.184}$$

$$\chi \to \varphi \qquad \text{2.185}$$

This is similar to the concept of negating the classifier. Under negation we have the mapping

$$\varphi \to 1 - \varphi \qquad \text{2.186}$$

$$\chi \to 1 - \chi \qquad \text{2.187}$$

Negation effectively negates the performance probabilities. We can create a negated classifier by simply doing the opposite of what the classifier tells us.

However, the mappings in 2.184-2.185 are fundamentally different. These mappings change TP into FP and FN into TN. This mapping is not realizable by simply manipulating the output of the classifier. In effect, this is an entirely different classifier.

We call two classifiers related by the mappings 2.184-2.185 duals. Alternatively, given a classifier with performance metrics φ_1 and χ_1, the its dual is a classifier with metrics

$$\varphi_2 \to \chi_1 \qquad \text{2.188}$$

$$\chi_2 \to \varphi_1 \qquad \text{2.189}$$

We can also negate the dual. The operations of negation and the dual relate four different classifiers together. These relations are shown in Table 24. In this table, if the original classifier has metric φ, then the negated metric will have $\varphi_{Neg.} = 1 - \varphi$, the dual has $\varphi_{Neg.} = \chi$, and the negated dual has the metric $\varphi_{Neg.} = 1 - \chi$.

Original	Negated	Dual	Negated Dual
φ	$1 - \varphi$	χ	$1 - \chi$
χ	$1 - \chi$	φ	$1 - \varphi$

Table 24: Relations between an original classifier, its dual, the negation, and the negated dual.

Figure 19, Figure 20, and Figure 21 are graphical representations of the operations of negation, dual, and the negated dual. In each, there is an interchange between the original performance metrics and the metrics after performing the operation. The four classifiers (original, negated, dual, negated dual) represent the possible classifiers that can be created from two performance values.

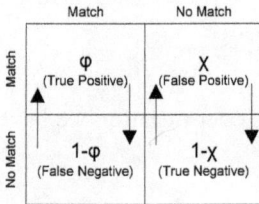

Figure 19: Negation.   Figure 20: Dual.   Figure 21: Negated Dual.

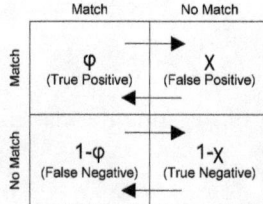

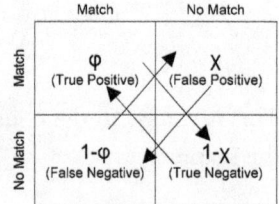

Metric	Negated	Dual	Negated Dual
TP	FN	FP	TN
FP	TN	TP	FN
FN	TP	TN	FP
TN	FP	FN	TP
AP	AP	AP	AP
AN	AN	AN	AN
CP	CN	CP	CN
CN	CP	CN	CP
TPR	FNR	FPR	TNR
FPR	TNR	TRP	FNR
ACC	-	ACC	-
TNR	FPR	FNR	TPR
FNR	TPN	TNR	FPR
PPV	NDR	FDR	NPV
NPV	FDR	NDR	PPV
FDR	NPV	PPV	NDR
NDR	PPV	NPV	FDR
MCC	-	-MCC	-
F1	-	-	-

Table 25: This table presents the relationships between the performance metrics for the four possible classifiers given two performance values in the case $\rho = .5$.

The dual changes TP into FP. Where a performance metric is dependent on ρ, the mapping between φ and χ will not map directly to another performance metric because of the asymmetry of ρ in the performance metrics (ρ attaches to TP and FN while $(1 - \rho)$ attaches to FP and TN). However, in the specific case where $\rho = .5$, then $\rho = 1 - \rho$, and the symmetry is restored[12].

Table 25 shows the relationships between the performance metrics in the case where $\rho = .5$. Although these relationships are not in general true, seeing how they are related in this specific case aids in understanding how the performance metrics are related in a conceptual sense.

The operations of negation and duals commute. Define the negation operator as

$$\mathcal{N}f(\varphi, \chi) = f(1 - \varphi, 1 - \chi) \qquad 2.190$$

Similarly, the dual operator is

$$\mathcal{D}f(\varphi, \chi) = f(\chi, \varphi) \qquad 2.191$$

Examine the result of acting the negation, followed by the dual:

$$\mathcal{D}\mathcal{N}f(\varphi, \chi) = \mathcal{D}f(1 - \varphi, 1 - \chi) \qquad 2.192$$

$$= f(1 - \chi, 1 - \varphi) \qquad 2.193$$

Alternatively, if we take the dual first then negate,

$$\mathcal{N}\mathcal{D}f(\varphi, \chi) = \mathcal{N}f(\chi, \varphi) \qquad 2.194$$

$$= f(1 - \chi, 1 - \varphi) \qquad 2.195$$

The order of operations does not matter. We have

$$\mathcal{D}\mathcal{N}f(\varphi, \chi) = \mathcal{N}\mathcal{D}f(\varphi, \chi) \qquad 2.196$$

So the negation and dual operators commute.

---

[12] When $\rho = .5$, the input set has the same number inputs that truly belong to the category as those that do not.

2.9 Types of Binary Classifiers

2.9-a αB Classifiers

The αB classifier is the standard basic binary classifier. This classifier examines an input and determines whether the input does or does not belong to the category.

The analysis of the previous sections is consistent with the αB classifier. As we will see in the next sections, changing the understanding of the nature of the classifier alters the interpretation of the performance metrics.

2.9-b βB Classifiers

The βB classifier is a binary classifier that places inputs in one of two categories. This is fundamentally different than the αB classifier. The αB places an input in a category or not. The βB classifier places the input in one of two categories.

At first, this may seem to be a semantic difference. However, the βB predetermines that there can only be two possible categories for each input. For example, we can make a classifier that examines a coin toss and determines if a given toss comes up heads. The αB would have the category 'heads', and each input would be labeled either 'heads' or 'not heads'.

A similar βB classifier would have the categories 'heads' and 'tails'. Each input would be placed into one of these two categories, and there is no possibility for an input to be in any other category. In a typical implementation, this βB operates just like the αB, only the βB classifier interprets 'not heads' as 'tails'.

For most inputs, interpreting 'not heads' as 'tails' is correct. However, there is a chance, however small, that the tossed coin will land on edge. The αB classifier might correctly place this input in the category of 'not heads', while the βB classifier might incorrectly interpret this as 'tails'.

In addition, one category's false positive becomes the other categories false negative. If a βB incorrectly identifies a true 'heads' as 'tails', this is a false negative for the 'heads' category. However, this is also a false positive for the 'tails' category.

The mixing of false positives and false negatives can lead to confusion in interpreting the performance results of the classifier. When examining a βB classifier, we need to be very careful in how we define the performance results in order to fully understand what the metrics tell us.

2.9-c γB CLASSIFIERS

The γB classifier places inputs into a category or otherwise is undetermined. This classifier recognizes that the undetermined category has inputs that do in fact belong in the category; however, the classifier is unable to detect this.

Again, this may seem at first to be a semantic difference with the αB classifier. However, in this case, false negatives are not interpreted in the same manner. Because we expect the γB classifier to intentionally place inputs into the undetermined category, a false negative is not necessarily a failure of the classifier.

The γB classifier acts more like a filter than a true classifier. γB classifiers are often designed to have a very high positive predictive value, and this metric may be more important than the other related metrics. In this case, we want to be certain that whatever the classifier identifies as in the category is truly in the category.

γB are very useful when the input set is heavily weighted toward inputs that are not in the category. For example, suppose we have a medical diagnostic that have a 99% true positive rate and a 99% true negative rate. At first, this seems like a very effective classifier. However, suppose that we are trying to detect a very rare disease which affects only 1 in 10,000 people. If we test one million people, we expect only 100 people will actually have the disease. The classifier will correctly identify 99 of those people. However, the classifier will incorrectly identify 1%, roughly 9,900 people, as having the disease when they do not (the true negative rate is 99%, and there are 990,000 people who do not have the disease in the input set). In this case, the classifier will identify a total of 9,999 people as having the disease, and only 1% (99 people) of those testing positive will actually be positive.

Cases such as the example naturally lead to γB classifiers. In these classifiers, the PPV metric dominates the performance characteristics, and examination of metrics such as the true/false positive rate is of secondary consideration.

3 ROC Analysis

3.1 ROC Space

The Receiver Operating Characteristic (ROC) is a statistical method used to graphically display the performance metrics of binary classifiers. From the previous chapter, we know that the performance of a binary classifier can be characterized with just two performance metrics. ROC space is the space determined by the potential values of the TP frequency (φ) and the FP frequency (χ).

Graphically, ROC space is the unit square formed from the region $x \in [0,1]$ and $y \in [0,1]$. φ (TP rate) is plotted on the y-axis, while χ (FP rate) is on the x-axis.

Figure 22 presents a simple example of a ROC plot. In this example we have a classifier where $\varphi = .75$ and $\chi = .25$. In the graph, the dashed lines provide the boundaries of the ROC space. All points must lie in the region $\varphi \in [0,1]$, $\chi \in [0,1]$.

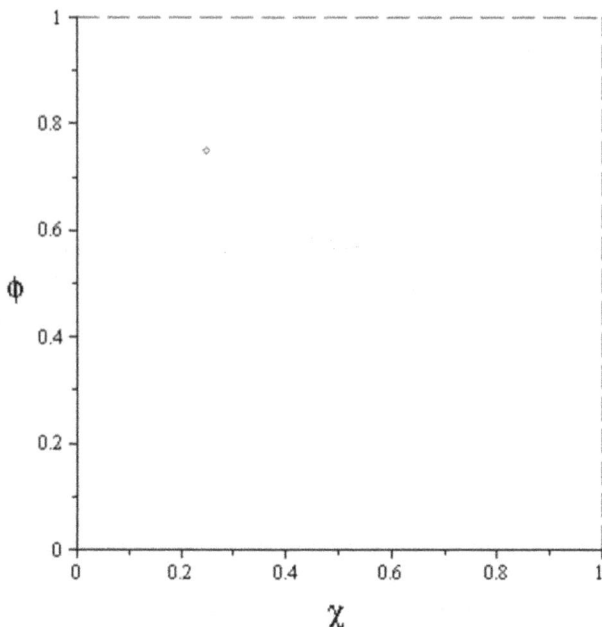

Figure 22: Simple ROC plot for a binary classifier with $\varphi = .75$ and $\chi = .25$.

We can also add error bounds to the graph. Figure 23 shows an example of a performance point with error bounds indicated. The inner region is a 95% confidence interval, while the outer region is a 99% confidence interval.

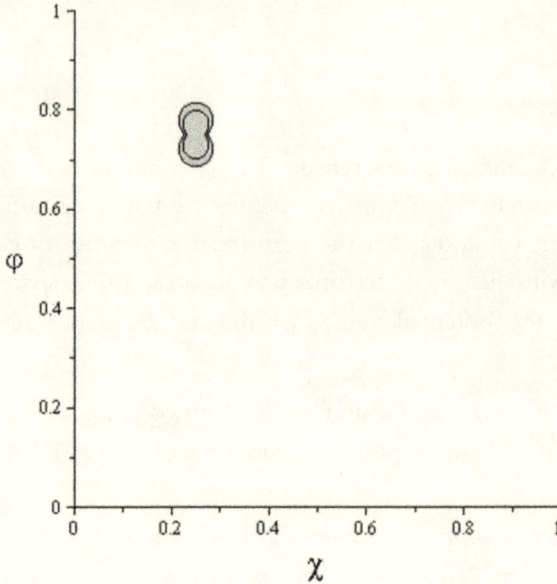

Figure 23: Simple ROC plot with error bounds for a binary classifier with $= .75$, $\chi = .25$, $\sigma_x = .01$, and $\sigma_y = .025$. The inner shaded region are points within the 95% confidence level, while the outer shaded region are points within the 99% confidence level.

Error bounds on points in ROC space are discussed in more detail in section 2.7. Here, it is important to note that the shaded regions surrounding a performance point indicate the region where the confidence level is less than some threshold. Typically, we will show the 95% and 99% confidence levels.

Moreover, if two points have overlapping error regions, this does not necessarily mean that the points are within a given error tolerance. In order to determine if two points are inside or outside a specific error tolerance, we must compute the z-score from equation 2.134. The ROC plot should be used as a general informational tool, but it does not have sufficient quantifiable information to determine if two points are significantly separated.

Plots in ROC space are useful graphical representations of the performance of a binary classifier. Multiple classifiers may be plotted together to indicate the relative performance of the classifiers.

In the following sections we examine some of the extremes of binary classifiers and how these look in ROC space. Understanding these examples provides a foundation for examining ROC plots in general.

3.1-a IDEAL CLASSIFIER

In the ideal case, the TP rate is one while the FP rate is zero. In this case, $\varphi = 1$ and $\chi = 1$. This point is the upper left corner of the ROC space. Figure 24 indicates this point on a ROC graph.

In addition, we may also be interested in classifiers where $\varphi = 1$ but $\chi \neq 1$. These classifiers all have the TP rate at 1, but may have non-zero FP rate. Figure 25 shows the points in ROC space where $\varphi = 1$. These points occupy the top boundary of the ROC space.

The line in Figure 25 demonstrates the extreme TP rate in ROC space. Classifiers here have perfect TP rate. As we move vertically away from this line, performance decreases with respect to the classifier's ability to properly identify inputs that belong to the category.

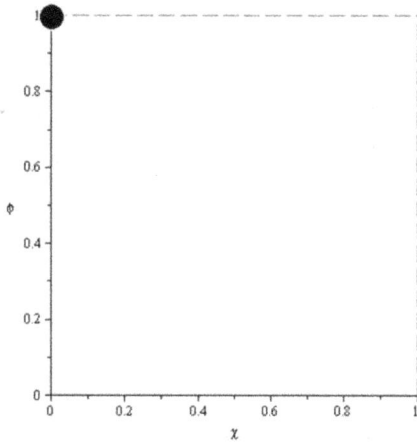

Figure 24: ROC graph for the ideal classifier.   Figure 25: ROC graph where $\varphi = 1$.

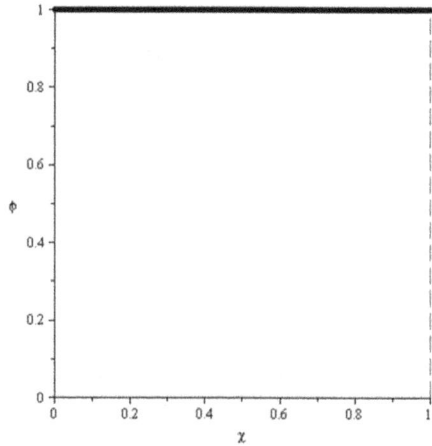

3.1-b FALLACIOUS CLASSIFIER

The fallacious classifier is always wrong. Here, the TP rate is zero while the FP rate is one. This classifier is plotted at the point $\varphi = 0$ and $\chi = 1$. This point is the lower right corner of the ROC space as shown in Figure 26.

Similarly, the line $\chi = 1$ represents all classifiers where the FP rate is one. These points occupy the right boundary of the ROC space as shown in Figure 27.

Comparing with Figure 25, we see that the line from Figure 27 meets the line in Figure 25 at the point at the upper right corner. At this point, the classifier demonstrates both a perfect TP rate and the worst possible FP rate. A classifier at this point is always right when it says that an input belongs to the class, but is always wrong when it says the input is not in the class.

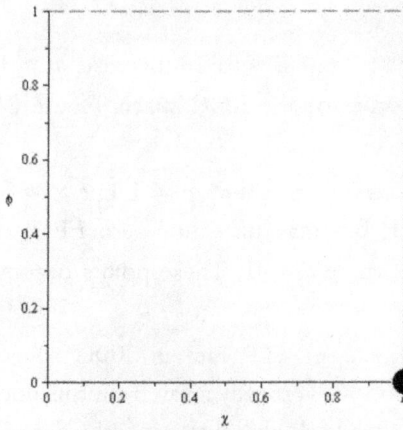

Figure 26: ROC graph for the fallacious classifier.

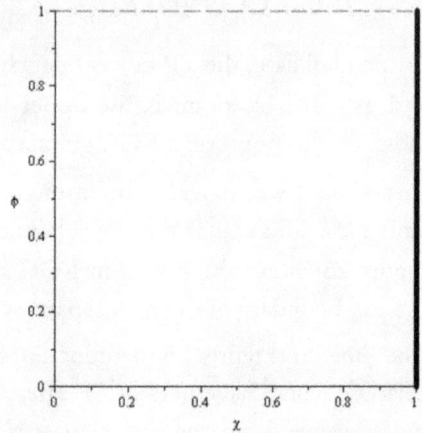

Figure 27: ROC graph where $\chi = 1$.

3.1-c RANDOM CLASSIFIER

The random classifier randomly assigns each input to the category. The metrics for the random classifier are provided in Table 13. Using 2.39 and 2.40 we find the rates as

$$\phi = \rho \qquad\qquad 3.1$$

$$\chi = \rho \qquad\qquad 3.2$$

In ROC space, this is the equation of the line

$$\phi = \chi \qquad\qquad 3.3$$

The ROC space for a random classifier with $\rho = .5$ is shown in Figure 28. The metrics for this classifier are at the center of the ROC space. From equation 3.3, we see that random classifiers have metrics along the line $\varphi = \chi$. This line is shown in Figure 29.

The line in Figure 29 is called the chance diagonal. Any classifier whose metrics lie on this line is effectively equivalent to a random classifier under this input set. It is important to remember that the metrics are tied to a particular input set. A classifier that is equivalent to a random classifier on one input set may not be equivalent on another input set.

We should keep in mind that a random classifier applied to a specific input set may have any performance metrics. A random classifier may by chance correctly identify every input and thus have ideal metrics (point in the upper left corner).

The chance diagonal represents the expectation of the metrics for a random classifier.

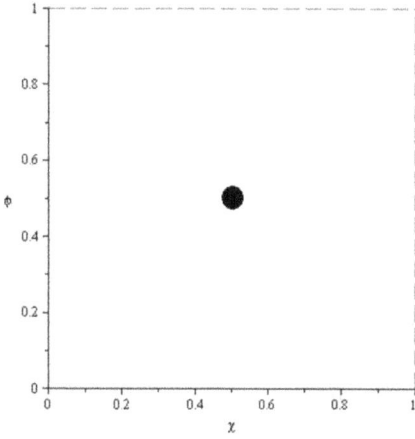

Figure 28: ROC graph for a random classifier with $\rho = .5$.

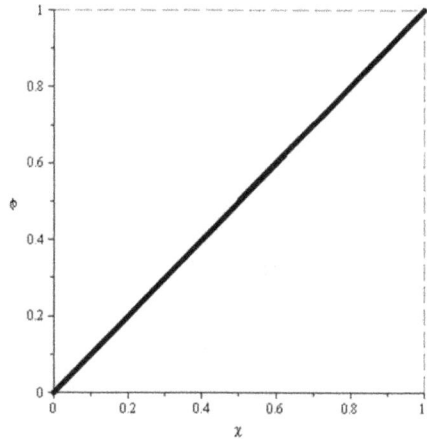

Figure 29: ROC graph for all possible random classifiers.

3.1-d NEGATION

Negation of a classifier transforms the metrics of the classifier as

$$\varphi \rightarrow 1 - \varphi \qquad \qquad 3.4$$

$$\chi \rightarrow 1 - \chi \qquad \qquad 3.5$$

In ROC space, this transformation reflects the performance point though the point (.5,.5). Given the performance point for a classifier, we reflect the point through the center of the ROC space. The reflected point is the performance point of the negated classifier.

The negated classifier is always achievable by simply doing the opposite of what the classifier tells us. If the classifier determines that an input is in the category, the negated classifier places the input not in the category. Similarly, if the classifier places an input not in the category, the negated classifier puts the same input in the category.

The relationship between a classifier and its negation is shown in Figure 30. The point in the center of the graph is the point of reflection for negation. If we negate a classifier that has performance metrics at this point, then the negated classifier has the same performance metrics.

Alternatively, a classifier with the performance metrics (0,1) transforms to (1,0). This relates the classifier positioned in the upper left corner to the negation in the lower right corner.

In general, we take the performance metrics for a classifier and reflect these values through the point in the center of the ROC space.

3.1-e DUAL

The dual of a classifier transforms the metrics of the classifier as

$$\varphi \to 1 - \varphi \qquad\qquad 3.6$$

$$\chi \to 1 - \chi \qquad\qquad 3.7$$

If we plot this transformation in ROC space, we see that the performance metrics are reflected about the diagonal $\varphi = \chi$. Given a performance point, we draw a line from the point to the diagonal, then find the equivalent point along the line on the other side of the diagonal.

The dual is not achievable based on the results of the classifier. There is no way to manipulate the results of a classifier to create its dual. The dual interchanges the TP and FP rates. These rates are based on different inputs. Thus, there is no way to create the dual from the original classifier.

Figure 31 shows the relationship between a classifier and its dual. We see that the point (1,0) has dual (0,1). This is the same result as for negation. For this point, the dual and the negation are the same. However, in general, the performance metrics are reflected about the diagonal, and the dual is distinct from the negation.

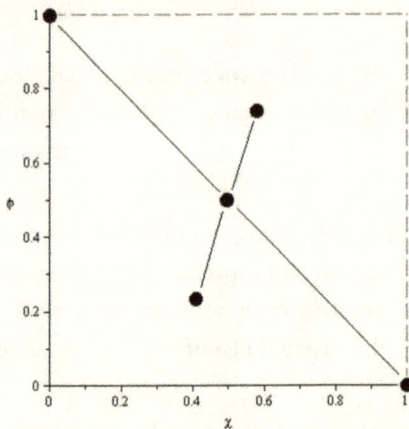

Figure 30: Negation of a classifier reflects the performance point in ROC space through the point (.5,.5).

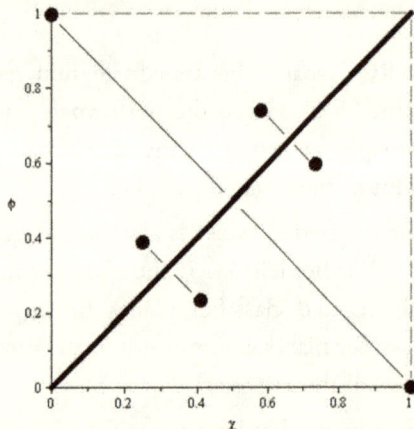

Figure 31: The dual of a classifier reflects the performance point in ROC space about the diagonal $\varphi = \chi$.

3.1-f NEGATED DUAL

We can also examine the negation of the dual. The negated dual transforms the original performance metrics as

$$\varphi \rightarrow 1 - \chi \qquad\qquad 3.8$$

$$\chi \rightarrow 1 - \varphi \qquad\qquad 3.9$$

The negated dual is the reflection of the performance metrics about the line $\varphi = 1 - \chi$. The negated dual is the combination of applying the negation and the dual to the performance metrics. Effectively, this reflects the performance metrics about the opposite diagonal as the dual.

The negated dual is not reachable based on the original performance metrics for the same reason that we cannot achieve the dual. Again, the negated dual related the TP rate with the FP rate. Since these originate from different input sets, there is no way to manipulate the results of a classifier to create the negated dual.

Figure 32 provides examples of the negated dual in ROC space. Performance points are reflected through the negatively sloped diagonal. The point (1,0) lies on the diagonal, and maps to itself. Similarly, the point (0,1) also lies on the diagonal and it too maps to itself. Generally, points are reflected through the negative diagonal.

Figure 33 shows a sample point with its negation, dual, and negative dual. In general, these four points have two points lying above the chance diagonal and two points below.

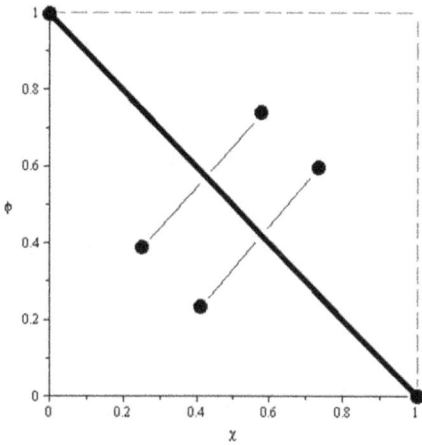

Figure 32: Negated Dual. First the point is reflected about the center (negation), then we reflect about the diagonal (dual).

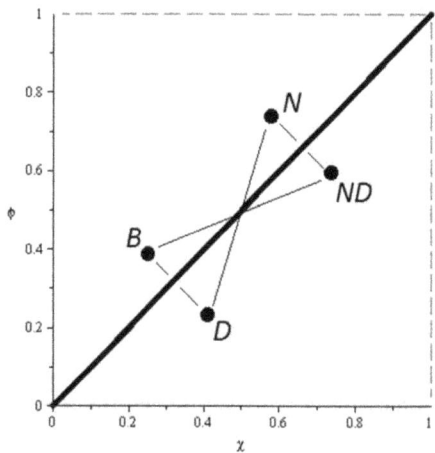

Figure 33: Example of an original classifier (B), its negation (N), dual (D), and negated dual (ND).

3.2 Probabilistic Classifiers

The binary classifiers we have examined so far result in a single set of performance metrics. We test the classifier against an input set, measure the TP rate and FP rate, and plot these in ROC space. In many cases we can extend this and have a single classifier produce a set of performance metrics.

There are two common methods used to assign a set of performance metrics to a single classifier. First, our classifier may have a tunable parameter. Such parameters are usually part of the underlying implementation of the classifier. For example, a classifier may examine an input, identify two measure values, and produce an output which is determined from a linear combination of the input values. In this case, the weighting between these two parameters is a continuous parameter that may be tuned. Different values of the parameter will result in different performance metrics for the classifier.

One of the most common methods used to assign multiple performance metrics to a classifier is through probabilistic and fuzzy algorithms. In each case, the classifier assigns an input a value on the range [0,1]. A value of 1 indicates that an input is definitely in the category, while a value of 0 indicates the input is definitely not in the category. Values between 0 and 1 indicate a probability or amount that the input belongs to the category.

We can create a tunable parameter from probabilistic and fuzzy methods by choosing a cutoff where we demark values that are assigned to the category versus values that we assign not in the category. For example, we might choose .75 as the cutoff. If the binary classifier determines the probability as greater than .75, we assign the input to the category. If the classifier determined the probability as .75 or less, we assign the input as not in the category.

For any given value of the cutoff, we can compute performance metrics for the classifier. By varying the cutoff from 0 to 1, we can obtain a set of performance metrics characterizing the classifier.

In either case, a tunable parameter allows us to assign a range of performance metrics to a binary classifier for a given input set. These performance values are often plotted in ROC space and generate a ROC curve for the classifier.

Let's illustrate this procedure with a simple example. Table 26 provides some sample data for 21 different points. This table lists the true value for each point (1 for in the class, 0 otherwise), and the output of our classifier.

We then set a value for the cutoff. When the classifier output is above the cutoff, the classifier places the input in the category. Table 27 provides the TP, FN, FP, and TN counts for cutoffs set at 0, .2, .4, .6, .8, and 1. From this, we can compute six different performance metrics, one for each cutoff value.

Input #	Output	True	Input #	Output	True	Input #	Output	True
1	0.15	0	8	.76	1	15	.38	1
2	.65	1	9	.62	0	16	.19	0
3	.98	1	10	.92	1	17	.82	0
4	.02	0	11	.85	1	18	.14	0
5	.37	0	12	.95	1	19	.69	1
6	.89	0	13	.12	0	20	.34	0
7	.47	1	14	.23	0	21	.99	1

Table 26: Sample table of inputs to a binary classifier (#), the output of the classifier (Output), and the true assignment to the class (True).

Cutoff	TP	FN	FP	TN
0.0	10	0	11	0
0.2	10	0	6	5
0.4	9	1	3	8
0.6	8	2	3	8
0.8	5	5	2	9
1.0	0	10	0	11

Table 27: TP, FN, FP, and TN counts for different values of the cutoff.

We can graph these results as the ROC plot shown in Figure 34. Alternatively, it is common to connect the performance points with lines to better indicate the performance of the classifier over continuous cutoff values. This is shown in Figure 35.

Figure 35 is a traditional ROC curve. With a tunable parameter, it is common for the curve to begin at the point (0,0), continue above the change diagonal, and end at (1,1).

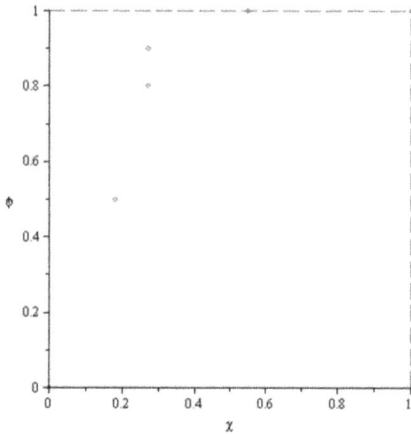

Figure 34: Performance points plotted in ROC space for the classifier from Table 27.

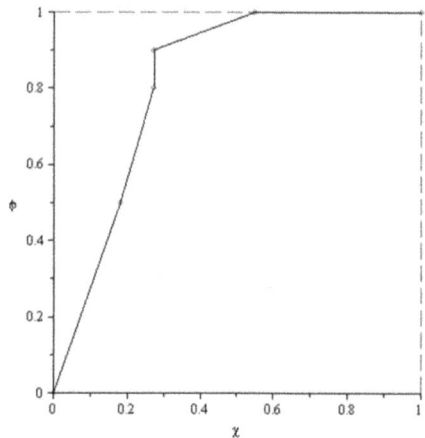

Figure 35: ROC curve joining the performance points from Figure 34.

ROC curves begin at (0,0) because at one extreme of the parameter, all inputs are assigned as not in the category. In this case the TP rate must be zero because no inputs are ever assigned to the class. In addition, the FP rate is also zero because there can be no false positives when no input is ever assigned to the class.

Alternatively, at the other extreme, the classifier often assigns all inputs to the category. In this case the TP rate must be one because there can be no false negatives when no inputs are assigned to the class. On the other hand, the FP rate is typically one because all inputs that should not be assigned to the class are placed into the class by the classifier. Hence, there are no TNs, so the FP rate is one.

3.3 Evaluating Classifier Performance

Generally, performance curves in ROC space are continuous curves moving from the origin to the point (1,1). ROC curves are typically above the chance diagonal because any region where the curve is below the diagonal, we can simply negate the classifier and obtain a performance point above the chance diagonal.

A ROC curve does not directly present an optimal value for the parameter. The curve represents a tradeoff between φ and χ. Unless the curve passes through the optimal point (0,1), we cannot determine an optimal value.

We must remember that the purpose of the classifier is to evaluate inputs that we do not already know the correct classification. We use a test input set where we have prior knowledge of the true values for the purpose of testing the classifier. Once the performance characteristics of the classifier have been evaluated, we usually desire to have the classifier work against inputs where we do not know the true value (in the class or not).

In applying our classifier to an unknown input set, the ROC curve demonstrates the various tradeoffs between φ and χ. When we compare two points with different values of φ and χ, we cannot always state that one point is better than another.

3.3-a COMPARABLE / INCOMPARABLE POINTS

Comparing two performance points in ROC space, when one performance point has both a higher value of φ and a lower value of χ, we can determine that this point is definitively better. However, when both the values of φ and χ are higher (or lower), we cannot definitively say that one is better than the other. These points are simply incomparable to each other.

Simply stated, when the line joining two performance points has negative slope, then one point is definitively better than the other. But when the line joining the points has positive slope, the points are incomparable. Figure 36 indicates the direction of lines connecting comparable points, while Figure 37 indicates slopes of lines joining incomparable points.

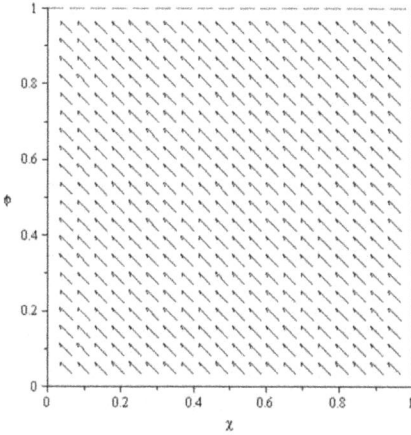

Figure 36: Comparable performance points in ROC space are joined by negatively sloped lines.

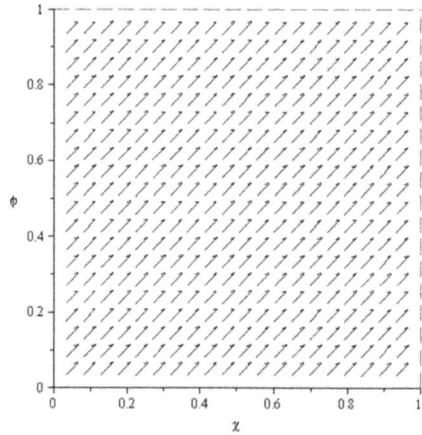

Figure 37: Incomparable performance points in ROC space are joined by positively sloped lines.

Incomparable points represent a tradeoff between the TP rate and the FP rate. These are points joined by positively sloped lines in ROC space. We cannot objectively choose between incomparable points. We need to examine the application of the classifier to the problem to determine which of these points represents the better solution to a given situation.

Moreover, we may determine that one point is more desirable when applying the classifier in one circumstance, while the other point is more desirable in other circumstances. Incomparable performance points present an opportunity for the user to tune the classifier to the problem at hand. In some cases, the tradeoff between φ and χ may lead us to choose higher φ even though this also increases χ. In other cases, we may come to the conclusion that the tradeoff is not worthwhile, and decide to take the lower χ even though this comes with a lower φ.

In any case, the tradeoff analysis must be done on a case-by-case basis. We need to examine how much the tradeoff is, the types of inputs, and the purpose of applying the classifier.

Alternatively, we can examine a cost function for the tradeoff between φ and χ (see § 3.4). If we have a function valuing φ in terms of χ, we can use this, coupled with the ROC curve, to determine an optimal value for the parameter.

3.3-b AUC

Given a performance curve in ROC space for a binary classifier, one way we can measure overall performance is by computing the Area Under the Curve (AUC). The AUC is the total area under the performance curve for the classifier in ROC space.

The minimum value for the AUC occurs when a classifier has performance metrics all along the χ-axis. Here, we find the AUC is zero. Alternatively, the maximum value for the AUC occurs when the performance curve is parallel to the χ-axis at $\varphi = 1$. In this case, the AUC is one. Generally, the AUC has value on the range [0,1].

$$AUC = \int_0^1 \varphi(\chi)\, d\chi \qquad\qquad 3.10$$

Moreover, since we can always use negation to move the performance metrics above the chance diagonal, the performance curve may always exceed this diagonal. The AUC for a random classifier is the area under its performance curve, which is a triangle with unit base and unit height. Thus, the AUC for the random classifier is .5.

$$Lift = \int_0^1 \varphi(\chi)\, d\chi - .5 \qquad\qquad 3.11$$

The AUC for the sample classifier from Figure 35 is shown in Figure 38. Alternatively, the lift of this classifier above the random classifier is the area under the performance curve of the classifier and above the chance diagonal. This area is illustrated in Figure 39.

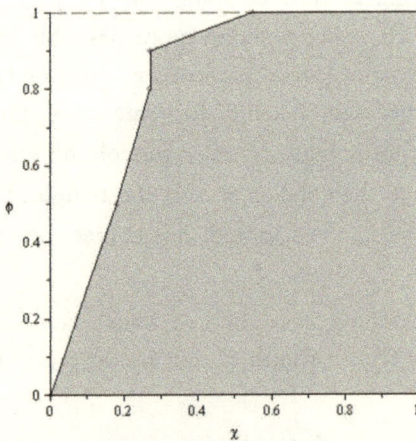

Figure 38: The AUC for the ROC curve from Figure 35.

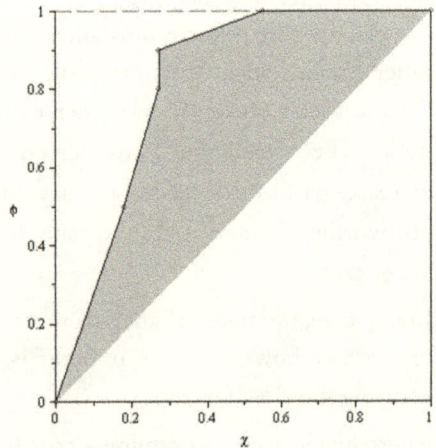

Figure 39: The lift above the chance diagonal for the ROC curve from Figure 38.

Similar to the lift, the Gini coefficient is the lift divided by the maximum value of the lift (.5):

$$g = \frac{AUC - .5}{.5} \qquad \text{3.12}$$

or,

$$AUC = 2g + 1 \qquad \text{3.13}$$

In some cases we may only have a partial performance curve. Under these circumstances we define the metric

$$PAUC = \int_a^b \varphi(\chi)\, d\chi \qquad \text{3.14}$$

This is the partial area under the curve. This metric is appropriate when the ROC curve is defined on the range [a,b].

Another common measure it the Youden Index. Its value is the maximum value of $\varphi - \chi$ over the performance curve. The Youden Index is on the range [-1,1]. As an example, the Youden Index for the classifier from Table 27 is .627 and occurs when the cutoff is .4.

Moreover, we can also compute the maximum vertical distance between the performance curve and the chance diagonal. However, since the chance diagonal is the curve $\varphi = \chi$, the MVD is equivalent to the Youden Index.

3.4 Minimizing Cost

In some cases we can examine two performance metrics and definitively determine when one set of metrics is superior to another. When the slope of the line connecting two points in ROC space is negative, we can identify one point as superior to the other.

When two points are connected by a positively sloped line, the performance metrics are incomparable. In general, we can have a set of points where each pair of points is connected via a positively sloped line. In such a situation, the points create a frontier of optimality.

In order to choose an optimal point on this frontier, we need to identify a cost or tradeoff function relating how to relatively value φ and χ. Generally, let $c(\varphi, \chi)$ be the cost function. Given a set of n incomparable points, we can simply substitute each value of φ and χ into the cost function to determine which has the least cost.

If we have a cost function and a finite set of performance metrics, we can find the optimum performance point by simply substituting in each pair of values for φ and χ to find the performance point that minimized the cost function.

Points that have the same value of the cost function are equivalent with respect to the cost. Such points form curves of equal cost in ROC space:

$$c(\varphi, \chi) = k \qquad\qquad 3.15$$

These equicost curves should have positive slope everywhere in ROC space. Positively sloped curves connect incomparable points. This means that the equicost curves are suitable tradeoff functions that may be used to evaluate incomparable points and determine an optimal solution.

If the equicost curves are negatively sloped, then the cost function connects comparable points. In this case, the cost function overrides the relationship between the TP and FP rates. Under these circumstances the ROC analysis is no longer valid. When the equicost curves are positively sloped, we can no longer rely on previous results.

For example, if our cost function is

$$c(\varphi, \chi) = 2\varphi + \chi \qquad\qquad 3.16$$

then we find the cost of the ideal classifier is $c(1,0) = 2$, while the cost of the fallacious classifier is $c(0,1) = 1$. Cost wise, the fallacious classifier is preferred over the ideal classifier! When the equicost curves have positive slope, the cost function overrides our understanding of the performance metrics of binary classifiers.

When the equicost curves have negative slope, the curves connect incomparable points. In this case, we can use the cost function to evaluate points that are otherwise incomparable in ROC space. For example, the cost function

$$c(\varphi, \chi) = -2\varphi + \chi \qquad\qquad 3.17$$

tells us that the cost of the point $(0,0)$ is $c(0,0) = 0$, while the point $(1,1)$ is $c(1,1) = -1$. These are both points on the chance diagonal, and are incomparable in ROC space. However, with the cost function, we see that the point $(1,1)$ is preferred to the point $(0,0)$.

As a more concrete example of cost functions, suppose that the binary classifier from Table 27 represents a classifier identifying defective parts on an assembly line. When the classifier identifies a part as 'defective', the part is discarded and the cost of the part is lost. When the classifier identifies a part as 'not defective', the part is used for manufacturing.

When a defective part is placed into manufacturing, the final product is defective. The products are under warranty, so a defective product is returned to the manufacturer. There is a cost associated with the processing returns and the repair and replacement of defective products.

Let c_p be the cost of the part, and let c_w be the cost associated with products returned under warranty. When the classifier determines that a part is good, and the part is actually good, then the product is valid and there is no cost. However, when the classifier determines that a part is good when the part is in fact defective, we incur a cost of c_w. When the classifier determines that the part is defective when in fact the part is not defective, we incur a cost of c_p.

The cost function may be expressed in terms of probabilities. Let $P(T_{\bar{A}}|C_A)$ be the probability that the part is defective when the classifier determines that it is not defective. Also, let $P(T_A|C_{\bar{A}})$ be the probability that the part is not defective when the classifier determines that it is defective.

Each of these must be weighted by the probability that the classifier identifies a part as good or defective. The probability the classifier identifies a part as good is

$$P(C_A) = \rho\varphi + (1-\rho)\chi \qquad 3.18$$

The probability the classifier identifies a part as defective is

$$P(C_{\bar{A}}) = \rho(1-\varphi) + (1-\rho)(1-\chi) \qquad 3.19$$

Putting everything together, the cost function is

$$c_w P(T_{\bar{A}}|C_A)P(C_A) + c_p P(T_A|C_{\bar{A}})P(C_{\bar{A}}) \qquad 3.20$$

Comparing this with Table 11 and Table 18, we can write the cost function in terms of φ and χ. First, from Table 11, we recognize that $P(T_{\bar{A}}|C_A) = FDR$, and that $P(T_A|C_{\bar{A}}) = NDR$. Furthermore, Table 18 provides expressions for these in terms of φ and χ. The cost function is

$$c_w \frac{\chi}{\chi + \frac{\rho}{1-\rho}\varphi}[\rho\varphi + (1-\rho)\chi] + c_p \frac{1-\varphi}{1-\varphi + (\rho^{-1}-1)(1-\chi)}[\rho(1-\varphi) + (1-\rho)(1-\chi)] \qquad 3.21$$

or,

$$c_w(1-\rho)\chi + c_p\rho(1-\varphi) \qquad 3.22$$

The equicost curves are given by

$$c_w(1-\rho)\chi + c_p\rho(1-\varphi) = k \qquad 3.23$$

or,

Substituting the values $\rho = .5$, with $c_w = 2$ and $c_p = 1$,

$$\chi + \frac{(1 - \varphi)}{2} = k \qquad\qquad 3.24$$

This forms a family of equicost curves. Figure 40 shows a few of these equicost curves. We see that the equicost curves have positive slope at every point.

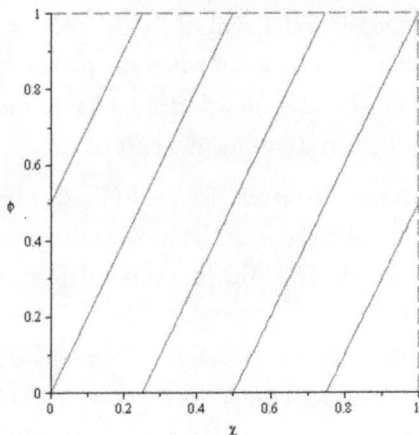

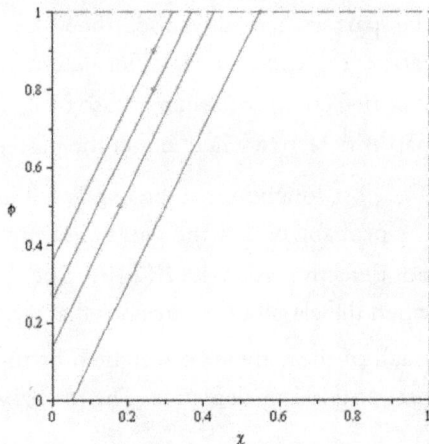

Figure 40: Equicost curves for the cost function in equation 3.24.

Figure 41: ROC curve joining the performance points from Figure 34.

We can apply this to the classifier metrics from Table 27. The cost for each cutoff is shown in Table 28. The extreme values of the classifier have infinite cost. The minimum cost occurs when the cutoff is set at .4.

The equicost curves at these points is shown in Figure 41. Each equicost curve is a positively sloped line.

For the parameters we set in equation 3.24, the optimal cutoff is at the value .4. However, choosing different parameters can lead to different optima.

Cutoff	φ	χ	k
0	1	1	1
0.2	1	.545	.595
0.4	.9	.273	.373
0.6	.8	.273	.487
0.8	.5	.5	.5
1	0	0	.5

Table 28: Cost table for the classifier metrics from Table 27.

3.5 Precision-Recall Curves

The TP and FP rates are enough to completely measure the performance of a classifier. These rates tell us how accurate our classifier is when presented with an actual positive (TP rate) or an actual negative (FP rate).

The TP rate is the frequency the classifier is correct when presented with an actual positive value. Specifically, the TP rate is the probability that the classifier is correct given an input is actually in the category:

$$TPR = P(C_A|T_A) = \frac{TP}{TP + FN} = \varphi \qquad 3.25$$

Alternatively, the FP rate is the frequency the classifier is incorrect when presented with an actual negative input. This rate is the probability that the classifier is wrong given an input is actually not in the category:

$$FPR = P(C_A|T_{\bar{A}}) = \frac{FP}{FP + TN} = \chi \qquad 3.26$$

In the above, C_A is the event that the classifier determines an input is in the category, T_A is the event that an input is actually in the category, and $T_{\bar{A}}$ is the event that the input is actually not in the category.

Although these metrics are sufficient to completely understand the performance of a classifier, the other metrics provide additional insight to classifier performance. Precision (PPV) and Recall (TPR) are also commonly used.

To understand the value of these metrics, examine the example from § 2.9-c. In this example, we have a classifier with a 99% TP rate and a 1% FP rate. However, because the input set is heavily weighted toward negative inputs, only 1% of the inputs that the classifier identifies as in the category are actually in the category.

From this example, we see that even though the classifier has overall good performance characteristics, the inputs that the classifier places into the category are only correct 1% of the time. In fact, we are asking a fundamentally different question here. We are asking 'What is the probability that the classifier is correct given that the classifier has determined that an input is in (or not in) the category?'.

Answering these questions is the purpose of the PPV. The positive predictive value (PPV) is the probability that the input is actually in the category given that the classifier has determined that the input belongs to the category:

$$PPV = P(T_A|C_A) = \frac{TP}{TP + FP} = \frac{\varphi}{\varphi + (\rho^{-1} - 1)\chi} \qquad 3.27$$

Alternatively, the TPR is simply the φ value we have examined in the previous sections:

$$TPR = P(C_A|T_A) = \frac{TP}{TP + FN} = \varphi \qquad 3.28$$

The PPV here is often called precision, while the TPR is referred to as recall. Precision measures how accurate the classifier is when the classifier determines that an input belongs to the category. Recall measures the overall ability of the classifier to identify an input that belongs to the category.

We can create a space in terms of precision and recall similar to the ROC space. We can plot performance points in the space defined by precision and recall rather than the $\varphi - \chi$ axes used in ROC space.

Let the variable $\theta = PPV$. θ is on the range [0,1]. The $\theta\varphi$ space is the unit square just like the $\chi\varphi$ (ROC) space. The ideal performance point in ROC space, $(\chi, \varphi) = (0,1)$, is the point $(\theta, \varphi) = (1,1)$ in $\theta\varphi$ space. Similarly, the fallacious classifier which is $(\chi, \varphi) = (1,0)$ in ROC space, becomes $(\theta, \varphi) = (0,0)$ in $\theta\varphi$ space.

The chance diagonal in ROC space is the line where $\varphi = \chi$. In $\theta\varphi$ space, these are the points $(\theta, \varphi) = (\rho, \varphi)$. These are vertical lines with intercept ρ. In $\theta\varphi$ space, a random classifier has performance metrics along a vertical line, and the placement of the line depends on the proportion of test inputs that belong to the class (ρ).

Examples of the chance diagonal in $\theta\varphi$ are provided in Figure 42. In addition, the performance points from Table 27 are show in Figure 43.

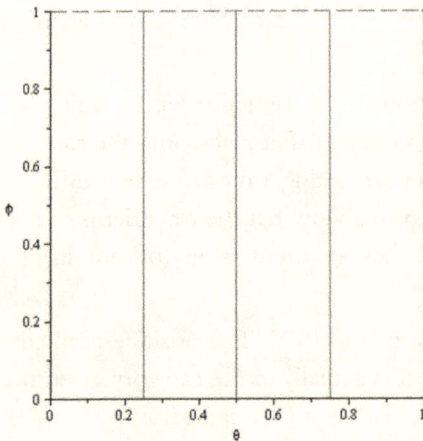

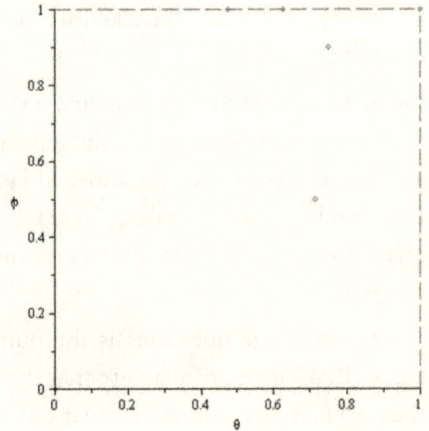

Figure 42: Random performance lines at the points $\rho = .25$, $\rho = .5$, and $\rho = .75$.

Figure 43: Points in $\theta\varphi$ space from Table 27.

4 Pitfalls with Binary Classifiers

4.1 Effect of the Input Set

When we test a binary classifier, we use an input set where we know if each input does or does not belong to the category. Let N be the total number of inputs in the test set, let ρN be the number of inputs in the test set that belong to the category, and let $(1 - \rho)N$ be the number of inputs in the test set that do not belong to the category.

We present each input to the classifier and tally the TP, FN, FP, and TN rates. Let P be the total number of actual positive results in the input set (ρN), and let F be the total number of actual negative results in the input set $((1 - \rho)N)$. In terms of the classifier tallies,

$$P = TP + FN \qquad\qquad 4.1$$

$$F = FP + TN \qquad\qquad 4.2$$

Let the variables

$$\phi = \frac{TP}{P} \qquad\qquad 4.3$$

$$\chi = \frac{FP}{F} \qquad\qquad 4.4$$

which makes

$$TP = \phi P \qquad\qquad 4.5$$

$$FP = \chi F \qquad\qquad 4.6$$

$$FN = (1 - \phi)P \qquad\qquad 4.7$$

$$TN = (1 - \chi)F \qquad\qquad 4.8$$

These variables are the variables used in examining ROC space from the previous chapter. We use these variables in the following sections to understand how the choice of input sets affects the performance metrics of the classifier.

4.1-a UNREPRESENTATIVE INPUT SET

The performance metrics of a classifier are tied to the test input set. The purpose of assessing the performance metrics is to determine how the classifier will perform when applied to a particular problem. However, if the test input set has different characteristics than the inputs from the problem, the classifier performance on the problem may be different than its performance on the test input set.

For example, suppose we have a classifier that analyzes documents and determines if the documents are written in English. We construct a test input set where a proportion ρ of the documents are in English, and the remainder, $1 - \rho$, are not in English. We present the test inputs to the classifier and find performance metrics φ and χ.

Suppose we then apply the classifier to a set of documents where the proportion of English documents is different than ρ. Will our classifier perform the same?

Let $\hat{\rho}$ be the proportion of English documents for the application of the classifier. We computed the performance metrics for the classifier in terms of φ and χ in § 2.6. We reproduce the results in Table 29 here for convenience.

Metric	Formula	Metric	Formula
TP	$\rho\varphi N$	FP	$(1 - \rho)\chi N$
FN	$\rho(1 - \varphi)N$	TN	$(1 - \rho)(1 - \chi)N$
AP	ρN	AN	$(1 - \rho)N$
CP	$[\rho\varphi + (1 - \rho)\chi]N$	CN	$[\rho(1 - \varphi) + (1 - \rho)(1 - \chi)]N$
TPR	φ	PPV	$\dfrac{\varphi}{\varphi + (\rho^{-1} - 1)\chi}$
FPR	χ	NPV	$\dfrac{1 - \chi}{1 - \chi + \dfrac{\rho}{1 - \rho}(1 - \varphi)}$
ACC	$\rho\varphi + (1 - \rho)(1 - \chi)$	FDR	$\dfrac{\chi}{\chi + \dfrac{\rho}{1 - \rho}\varphi}$
TNR	$1 - \chi$	NDR	$\dfrac{1 - \varphi}{1 - \varphi + (\rho^{-1} - 1)(1 - \chi)}$
FNR	$1 - \varphi$	MCC	$(\varphi - \chi)\sqrt{\dfrac{\rho(1 - \rho)}{[\rho\varphi + (1 - \rho)\chi][\rho(1 - \varphi) + (1 - \rho)(1 - \chi)]}}$
F1	$\dfrac{2\varphi^2}{\varphi + \varphi^2 + (\rho^{-1} - 1)\chi}$		

Table 29: Performance measures in terms of φ and χ.

Examining Table 29, we see that most of the performance metrics depend on the value of ρ. Any of the performance metrics that depend on ρ will change when the proportion of the test input set differs from the problem inputs.

However, the base metrics, φ and χ, do not depend on ρ. For instance, φ is the proportion of actual positives that the classifier determines are positive. This proportion does not depend on ρ. If we use an input set where the proportion of actual positives is $\hat{\rho}$, the number of positives detected by the classifier will change, but we expect that the ratio between detected positives and true positives will remain the same. The same arguments hold for χ.

Based on this, we see that if the proportion of positives in the problem domain is different than the test inputs, many of our performance metrics are affected. In fact, only TPR, FPR, TNR, and FNR are not affected by a change in the proportion ρ of the input set.

Many of the performance metrics change values with the input set ratio changes. When we are comparing classifiers, is it possible that one classifier can outperform another with one input ratio, but then the other classifier performs better when the input ratio changes?

In order for the rank ordering to change, we must have the performance metrics between two points ordered one way when the input set has proportion ρ, but then ordered the opposite way when the input set has proportion $\hat{\rho}$. Let $f(\rho; \varphi, \chi)$ be a metric. For the rank ordering to change we must have

$$f(\rho; \varphi, \chi) < f(\rho; \bar{\varphi}, \bar{\chi}) \qquad \text{4.9}$$

and

$$f(\hat{\rho}; \varphi, \chi) > f(\hat{\rho}; \bar{\varphi}, \bar{\chi}) \qquad \text{4.10}$$

If we can find a ρ and $\hat{\rho}$ satisfying these expressions, then the metric is sensitive to changes in the input set proportions. Under these circumstances, the rank ordering we find with one input set may be different than the rank ordering we obtain using an input set with different proportions.

Examining the metrics from Table 29, four of the metrics are of the form

$$\frac{x}{x + g(\rho)y} \qquad \text{4.11}$$

We can test this expression using equations 4.9 and 4.10. First, from 4.9 we have

$$f(\rho; x, y) < f(\rho; \bar{x}, \bar{y}) \qquad \text{4.12}$$

$$\frac{x}{x + g(\rho)y} < \frac{\bar{x}}{\bar{x} + g(\rho)\bar{y}} \tag{4.13}$$

$$x(\bar{x} + g(\rho)\bar{y}) < \bar{x}(x + g(\rho)y) \tag{4.14}$$

$$g(\rho)x\bar{y} < g(\rho)\bar{x}y \tag{4.15}$$

$$x\bar{y} < \bar{x}y \tag{4.16}$$

We can find the result from 4.10 by using the last expression, substituting $\hat{\rho}$ for ρ, and changing the direction of the inequality. Thus,

$$x\bar{y} > \bar{x}y \tag{4.17}$$

Neither of these equations depends on $\hat{\rho}$ or ρ. Furthermore, there is no consistent solution to the two equations. This contradiction means that there is no variables $\varphi, \chi, \bar{\varphi}, \bar{\chi}$ such that changing from ρ to $\hat{\rho}$ will cause the relative ordering to change.

Alternatively, if we examine the F1 metric we find:

$$\frac{2\varphi^2}{\varphi + \varphi^2 + (\rho^{-1} - 1)\chi} < \frac{2\bar{\varphi}^2}{\bar{\varphi} + \bar{\varphi}^2 + (\rho^{-1} - 1)\bar{\chi}} \tag{4.18}$$

$$2\varphi^2[\bar{\varphi} + \bar{\varphi}^2 + (\rho^{-1} - 1)\bar{\chi}] < 2\bar{\varphi}^2[\varphi + \varphi^2 + (\rho^{-1} - 1)\chi] \tag{4.19}$$

$$2\varphi^2[\bar{\varphi} + \bar{\varphi}^2 + (\rho^{-1} - 1)\bar{\chi}] < 2\bar{\varphi}^2[\varphi + \varphi^2 + (\rho^{-1} - 1)\chi] \tag{4.20}$$

$$2\varphi^2\bar{\varphi} + 2\varphi^2(\rho^{-1} - 1)\bar{\chi} < 2\bar{\varphi}^2\varphi + 2\bar{\varphi}^2(\rho^{-1} - 1)\chi \tag{4.21}$$

$$\varphi^2\bar{\varphi} - \bar{\varphi}^2\varphi < \bar{\varphi}^2(\rho^{-1} - 1)\chi - \varphi^2(\rho^{-1} - 1)\bar{\chi} \tag{4.22}$$

$$\frac{\varphi\bar{\varphi}(\varphi - \bar{\varphi})}{\bar{\varphi}^2\chi - \varphi^2\bar{\chi}} < (\rho^{-1} - 1) \tag{4.23}$$

The other condition is

$$\frac{\varphi\bar{\varphi}(\varphi - \bar{\varphi})}{\bar{\varphi}^2\chi - \varphi^2\bar{\chi}} > (\hat{\rho}^{-1} - 1) \tag{4.24}$$

In this case, we can find values for ρ and $\hat{\rho}$ satisfying these conditions provided the ratio on the left has suitable values. From this we see that F1 is sensitive to changes in the input ratio. Under the right circumstances, the rank order of classifiers under one input set may present a different rank ordering when using an input set with a different input ratio.

For most of the performance metrics, if we fix φ and χ and let ρ vary, the relative ordering of performance is unchanged. However, ACC, MCC, and F1 do not share this property. The relative ordering of these metrics is sensitive to changes in the input ratio.

Table 30 indicates which performance metrics preserve relative ordering and which do not when the input ratio ρ changes.

Metric	Preserves Relative Ordering	Metric	Preserves Relative Ordering
TP	-	FP	-
FN	-	TN	-
AP	-	AN	-
CP	-	CN	-
TPR	Yes	PPV	Yes
FPR	Yes	NPV	Yes
ACC	No	FDR	Yes
TNR	Yes	NDR	Yes
FNR	Yes	MCC	No
F1	No		

Table 30: Performance measures that change relative ordering between classifiers when φ and χ are fixed and the input ratio ρ varies.

We should take care when using ACC, MCC, and F1 to rank order classifiers. If the input ratio ρ used to measure performance differs from the input ratio in the application of the classifier, the rank ordering of the classifiers can change. When using these metrics to rank order classifiers, it is important that the input ratio used to measure classifier performance matches the input ratio for the application of the classifiers.

4.1-b HIDDEN SUBCLASS

In the previous section we examined how altering the input ratio can affect the performance metrics for classifiers. Although many of the performance metrics change value when the input ratio is changed, most of the metrics at least preserve relative ordering. Thus, if one classifier is better than another under one input ratio, we expect that the same classifier will perform better when the input ratio is changed.

A more insidious problem with the input set is the presence of hidden subclasses. Continuing the example from the previous section, suppose we are examining a classifier that determines if a document is written in English. Further, suppose we have both short and long documents. Initially, we do not distinguish between short and long documents. In this sense there is a hidden subclass to the input set.

Further, suppose the performance of our classifier is different on short versus long documents. Let (χ_s, φ_s) be the performance metrics of the classifier on short documents, while (χ_l, φ_l) represents the performance metrics on long documents.

If our input set has a proportion ε_s of short documents and ε_l of long documents, our measured performance is expected to be

$$\varphi = \varepsilon_s \varphi_s + \varepsilon_l \varphi_l \qquad \text{4.25}$$

$$\chi = \varepsilon_s \chi_s + \varepsilon_l \chi_l \qquad \text{4.26}$$

Now suppose that when we apply the classifier, the proportions are $\hat{\varepsilon}_s$ and $\hat{\varepsilon}_l$. Our performance measures are expected to be

$$\varphi = \hat{\varepsilon}_s \varphi_s + \hat{\varepsilon}_l \varphi_l \qquad \text{4.27}$$

$$\chi = \hat{\varepsilon}_s \chi_s + \hat{\varepsilon}_l \chi_l \qquad \text{4.28}$$

Under these circumstances, the base performance metrics are different. For example, if out test input set is all long documents, our performance measures are expected to be $(\chi, \varphi) = (\chi_l, \varphi_l)$. If these problem inputs are all short documents, we expect the performance to be $(\chi, \varphi) = (\chi_s, \varphi_s)$. At these extremes, the performance metrics under the test inputs can be completely different from the performance metrics when we apply the classifier to a particular problem.

If we have two classifiers, where one outperforms another, the presence of a hidden subclass can change the relative ordering when we apply these classifiers to a particular problem. For example, suppose we have one classifier whose performance on short documents is $(0,1)$ (perfection) but performs as $(.5,.5)$ (random) on long documents. Suppose we have another classifier that performs as $(.5,.5)$ on long documents but has $(0,1)$ on short documents.

In this example, the first classifier has superior performance on the test input set, but the second classifier has better performance when applied to the particular problem. Here we see how hidden subclasses can affect the relative ordering of the classifiers.

If these base performance metrics change, then all of our derived metrics will change as well. Hidden subclasses can completely change the relative ordering of classifiers.

It is critical that the test inputs reflect the inputs seen in the problem domain. The performance metrics we measure for a classifier are tied to the input set. When we apply the classifier in a situation where the makeup of the inputs is significantly different than the test inputs, the realized performance of the classifier may be different than expected.

4.1-c ASYMMETRIC INPUT SET

Asymmetric inputs sets occur when the problem inputs have either $\rho \approx 0$ or $\rho \approx 1$. If we test the classifier with an evenly distributed input set ($\rho \approx .5$), the realized performance may be surprising.

We discussed such a situation in § 2.9-c. In this example we see how a classifier that initially appears to have a high performance at $(\chi, \varphi) = (.01, .99)$ end up with only 1% of the inputs classified as positive that are correct. Here, it is important to understand what we are trying to achieve with our classifier.

In many cases we want a classifier that correctly classifies each input. In terms of probability, we want to have a probability that the classifier is correct given the true value.

Alternatively, in other cases we may want to have the pool of items that the classifier determines are positive to be actually posiyive. In probability, this is the probability that the inputs are truly positive given the classifier has identified them as positive.

These two variants are superficially similar, but they are distinctly different. It is important to understand what we desire to achieve with the classifier in order to choose appropriate metrics.

In the first case, the base metrics (χ, φ) are suitable for measuring the performance of a classifier. In the second case, the PPV or NPV are more appropriate because these metrics tell us if proportion of the inputs that are classified as positive or negative is high.

4.2 Precision versus Recall

Precision (PPV) is used when we want to know if the inputs that the classifier identifies as positive are truly positive. Recall is the proportion of inputs that the classifier correctly identifies as positive are actually positive. In terms of our base metrics,

$$Precision = PPV = P(T_A|C_A) = \frac{\varphi}{\varphi + (\rho^{-1} - 1)\chi} \qquad 4.29$$

$$Recall = TPR = P(C_A|T_A) = \varphi \qquad 4.30$$

Precision and recall are often used in the field of information retrieval. A common problem is to present a software program with a set of criteria and ask the program to retrieve all relevant documents matching the criteria.

There may be a debate about whether precision or recall is the more important performance measure. Do we want to be certain that the results we get back are all relevant (precision) or do we want to know that we have all relevant results somewhere in the items retrieved?

Different situations will call for different weights on these performance measures. Generally, we need to identify a tradeoff function in order to rank order classifiers based on their performance metrics.

However, there are certain questions that can be addressed based on the definition of the performance metrics. For example, perfect precision can be achieved by simply identifying a single relevant document. Often finding a single exact match is easily achieved. If we can reliably identify one matching document, then designing a classifier to retrieve only this one document will achieve perfect precision. The precision is perfect because there are no false positives, so $\chi = 0$. When $\chi = 0$, equation 4.29 becomes 1.

Examining 4.30, we only have the variable φ. On the surface, there does not seem to be any way to manipulate the value of the recall. This has led some to say that precision is easy to achieve, while recall is hard.

However, equation 4.30 is not a definition. This is only the expression of recall in terms of our base performance metrics. The TPR is defined as

$$TPR = \frac{TP}{TP + FN} \qquad\qquad 4.31$$

We can achieve perfect recall if we have no false negatives. Remember, false negatives occur when an input is truly positive and we classify it as negative. In the information retrieval example, false negatives occur when we fail to retrieve a document that we should have retrieved.

We can obtain no false negatives simply by returning every document on every query. If every document is returned, there can be no false negatives because there are no negatives at all.

We see that both high precision and high recall are easily achieved. High precision is achieved by returning only one document that is an exact match. High recall is achieved by returning all documents. Neither of these results is difficult to obtain, and neither is likely to be what we are looking for in the classifier.

In order to compare the performance of two classifiers, we need to first have an input set that is representative of the problem. Second, we either need to accept that certain performance measures are incomparable or we need to identify a tradeoff function. If we accept that performance measures may not be comparable, there are no winners, only a frontier of optimal performers. Alternatively, if we identify a cost function, we need to specify why this cost function appropriately measures the tradeoff between precision and recall.

If we decide to use a cost function, the ideal cost function will not be sensitive to the input ratio[13]. If the cost function is sensitive to the input ratio, our rank ordering may be sensitive as well. In this case, if the test inputs are not representative of the problem domain, we may obtain one rank ordering based on the test inputs and cost function, while we would get a different ordering based on the problem inputs and cost function.

4.3 Two-Class Binary Classifiers

Another problem we may encounter is incorrectly assigning a binary classifier to two classes. Typically, we want to have one class for the classifier. The classifier examines each input and determines if the input does not does not belong to the class.

However, in many situations it is tempting to assign 'not in a class' as membership in another class. For example, if a classifier examines coin tosses and identifies if the coin came up 'heads', we may be tempted to assign 'not heads' to 'tails'.

Assigning 'not in the class' to another class can lead to unexpected results. Classifiers are often applied via software programs to a large amount of data. When classifiers examine millions, billions, or more inputs, then even rare events are almost certain to occur.

In the coin toss example, it is very unlikely that the coin will land on edge. However, if we toss enough coins, there is a finite probability that we will have some land on edge. If the classifier determines that these are 'not heads', we then assign them to 'tails'. Doing this will lead us to believe that the coin is biased in favor of tails.

Moreover, assigning 'not in a class' to another class presents missed opportunities. Rather than using one classifier to determine 'heads' versus 'tails', examine two classifiers, one identifying 'heads' versus 'not heads', and the other identifying 'tails' versus 'not tails'. In this case, if a coin were to land on edge, we

---

[13] See § 4.1-a.

might find 'not heads' and 'not tails'. Examining such events may provide an opportunity to discover that there is another category, 'on edge', that we overlooked when setting up the problem.

4.4 Binary versus Multiclass Classifiers

Another problem that may be encountered is interpreting a multiclass classifier as a binary classifier. We discuss multiclass classifiers in later chapters. For now, we only need to understand that a multiclass classifier has multiple classes that we can use to categorize an input.

Multiclass classifiers can look much like binary classifiers. For example, suppose we want to identify names in a list. We provide a name to search on, and a software program compares this name to the names on the list to determine which match. This is an information retrieval problem.

The test name may be compared with each name on the list. Each comparison is a binary classifier: does the test name belong to the same class as this name on the list? We may be tempted to treat this as a binary classifier and use the metrics we have developed.

However, this is really a multiclass classifier. Each individual name is its own class. Different names are different classes. Each individual name will have its own performance metrics, independent from other distinct names. We should take care to understand this classifier from a multiclass perspective and use the tools we develop in the next chapters.

5 Trinary Classifier

5.1 Overview

The trinary classifier has three different classes, rather than the two classes for the binary classifier. For simplicity, in this section we will designate these classes as class A, B, and C.

We proceed with a method similar to that employed with the binary classifier. A test input set is constructed, where the true classification of each input is known. Each input is then presented to the classifier, and we tally the results.

The test input set will have a predetermined number of inputs from class A, B, and C. Let N be the total number of inputs, and let N_A, N_B, and N_C be the number of test inputs from class A, B, and C respectively. We designate the proportions as

$$\rho_A = \frac{N_A}{N} \qquad\qquad 5.1$$

$$\rho_B = \frac{N_B}{N} \qquad\qquad 5.2$$

$$\rho_C = \frac{N_C}{N} \qquad\qquad 5.3$$

The proportions are constrained to add to one:

$$\rho_A + \rho_B + \rho_C = 1 \qquad\qquad 5.4$$

We use these variables in the following sections as we examine the trinary classifier. Just as some of the binary classifier metrics depend on the input ratio, the trinary classifier metrics will depend on these input ratios.

5.2 Metrics

In this section we examine metrics for the trinary classifier. We generalize the metrics of the binary classifier to the case of three classes. We begin by generalizing the confusion matrix. We identify the constraints on the confusion matrix to determine the number of parameters needed to characterize the trinary classifier. Based on these results, we identify metrics similar to those used for the binary classifier.

5.2-a Confusion Matrix

The binary classifier only had a single class. We could view the results as either in or not in the class. Alternatively, we could view this as two classes, one class for A, and the other class for not A.

Trinary classifiers are different. There are at least two distinct classes. The third class may be interpreted as 'none of the above', or this may be a separate class entirely. We treat these as three separate classes.

The confusion matrix for the trinary classifier is shown in Figure 44. This confusion matrix has three columns, one for the actual value A, B, and C. Similarly, there are three rows, one for each possible output for the classifier. In total, there are nine elements to the confusion matrix.

We tally the confusion matrix similarly as the binary classifier. We present the classifier with an input. Suppose the actual value of the input is A. If the classifier identifies this correctly, we add one to the cell AA, which is the True Positive A (TA). If the classifier incorrectly identifies this as B, we add one to the cell BA, which is a false B, true value A (FB^A). Similarly, if the classifier incorrectly identifies this as C, we add one to the cell CA, which is a false C, true value A (FC^A).

This confusion matrix does not have a concept of false negatives. In the binary classifier, a false negative is the case where we expected to get A but the classifier identified the input as not A. For the trinary classifier, if we expect to get A, and the classifier identifies the input as B or C, we call this a false B or false C, rather than calling these false negatives.

Actual Value

		A	B	C
Predicted Outcome	**A**	TA (True Positive A)	FA^B (False A-B)	FA^C (False A-C)
	B	FB^A (False B-A)	TB (True Positive B)	FB^C (False B-C)
	C	FC^A (False C-A)	FC^B (False C-B)	TC (True Positive C)

Figure 44: Confusion matrix for the trinary classifier.

The confusion matrix is a matrix of counts for each of the nine categories based on presenting the test input set to the classifier. We are interested in the rates as well. Each column of values in the confusion matrix must add to a predetermined value based on the proportion of that class in the test inputs. For example, the values of column A must satisfy

$$TA + FB^A + FC^A = N_A = \rho_A N \qquad 5.5$$

Similarly,

$$FA^B + TB + FC^B = N_B = \rho_B N \qquad 5.6$$

$$FA^C + FB^C + TA = N_C = \rho_C N \qquad 5.7$$

We can create a confusion matrix in terms of rates by dividing the values in each column by the sum of the column. First, let's show this process for the binary classifier. Figure 45 shows the confusion matrix for the binary classifier. Again, each column must add to the number of inputs in the test set matching the class. If we divide the value of each column by the sum of that column, we get the matrix in Figure 46. After division, the values in each column add to one.

We notice in Figure 46, once we divide each column entry by the sum of its column, each cell in the resulting matrix is related to our base performance metrics. For binary classifiers, the proportions φ and χ are the base performance variables we analyzed to characterize the behavior of the classifier. This process provides insight for the trinary classifier metrics.

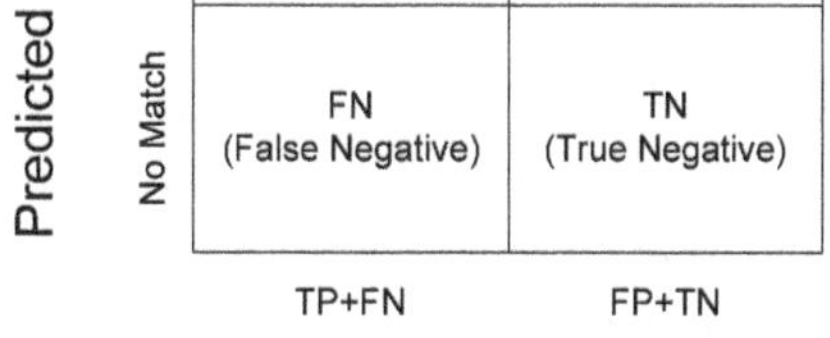

Figure 45: Confusion matrix for the binary classifier. The columns must add to the total number of actual inputs matching the class.

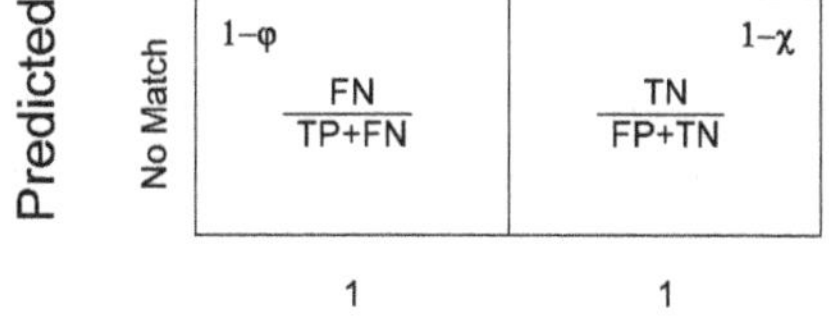

Figure 46: Normalized confusion matrix for the binary classifier. Each column adds to one.

Noting this, we can perform a similar procedure on the confusion matrix for the trinary classifier. We see from equations 5.5-5.7 that each column of the confusion matrix must add to the number of inputs in the test set that belong to each class.

The sums of the trinary confusion matrix are shown in Figure 47. If we divide each column entry by the sum of its column, we get the matrix in Figure 48. In Figure 48, we indicate variable names for each of the matrix cells. Because each column must add to one, on a given row we only need to specify two variables. The third variable is determined from the other two.

Looking back at the binary classifier, we see that the final row is written in terms of the other variables. We adopt a similar procedure for the trinary classifiers. We specify variables for each of the cells, except the cells in the final row. These cells we choose to write in terms of the other variables.

Actual Value (Figure 47)

Predicted Outcome	A	B	C
A	TA (True Positive A)	FA^B (False A-B)	FA^C (False A-C)
B	FB^A (False B-A)	TB (True Positive B)	FB^C (False B-C)
C	FC^A (False C-A)	FC^B (False C-B)	TC (True Positive C)
	$TA+FB^A+FC^A$	$FA^B+TB+FC^B$	FA^C+FB^C+TC

Actual Value (Figure 48)

Predicted Outcome	A	B	C
A	φ_A $\dfrac{TA}{TA+FB^A+FC^A}$	χ_A $\dfrac{FA^B}{FA^B+TB+FC^B}$	ψ_A $\dfrac{FA^C}{FA^C+FB^C+TC}$
B	φ_B $\dfrac{FB^A}{TA+FB^A+FC^A}$	χ_B $\dfrac{TB}{FA^B+TB+FC^B}$	ψ_B $\dfrac{FB^C}{FA^C+FB^C+TC}$
C	$1-\varphi_A-\varphi_B$ $\dfrac{FC^A}{TA+FB^A+FC^A}$	$1-\chi_A-\chi_B$ $\dfrac{FC^B}{FA^B+TB+FC^B}$	$1-\psi_A-\psi_B$ $\dfrac{TC}{FA^C+FB^C+TC}$
	1	1	1

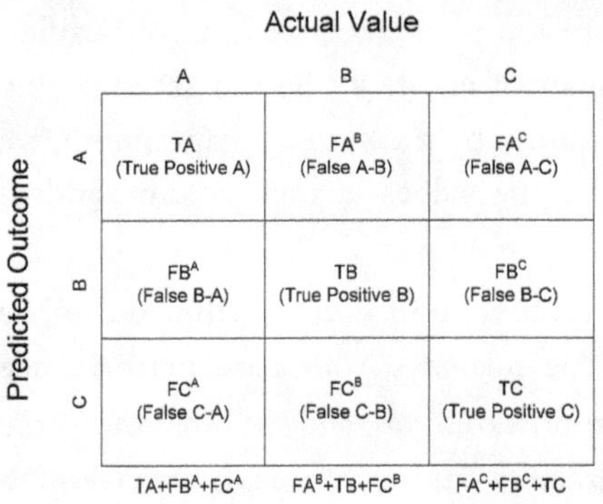

Figure 47: Confusion matrix for the trinary classifier with the sum of each column.

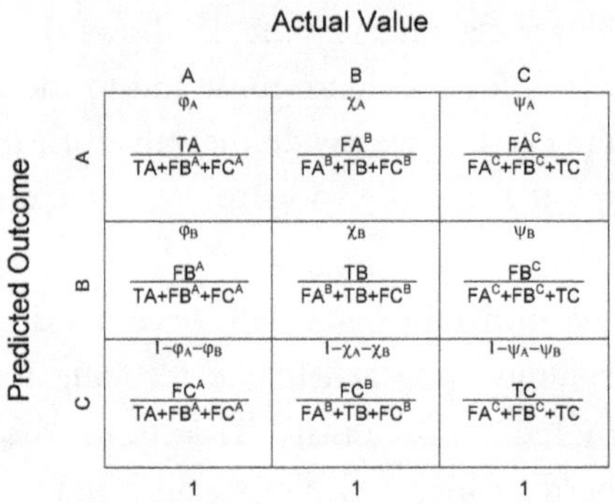

Figure 48: Normalized confusion matrix for the trinary classifier.

Each column in Figure 48 is constrained to add to one. Because of this, we are left with only six variables. We use these variables as the base performance metrics for the trinary classifier. Table 31 provides a list of these six variables and their expressions in terms of the tally from the trinary classifier confusion matrix.

In choosing these variables, we adopted a procedure consistent with that used in the binary classifier. However, there are many potential choices for the base variables. In any given column, all cells but one will be a base metric. The remaining cell is determined from the constraint that all cells in the column must add to one. Thus, although we choose to eliminate the final row, in fact, we could have eliminated any row. Alternatively, we could eliminate the diagonal and use the non-diagonal cells as the base metrics. All of these choices are effectively equivalent.

Variable	Expression
N	Number of inputs in the test set
ρ_A	Proportion of inputs in the test set that belong to class A
ρ_B	Proportion of inputs in the test set that belong to class B
φ_A	$\dfrac{TA}{TA + FB^A + FC^A}$
φ_B	$\dfrac{FB^A}{TA + FB^A + FC^A}$
χ_A	$\dfrac{FA^B}{FA^B + TB + FC^B}$
χ_B	$\dfrac{TB}{FA^B + TB + FC^B}$
ψ_A	$\dfrac{FA^C}{FA^C + FB^C + TC}$
ψ_B	$\dfrac{FB^C}{FA^C + FB^C + TC}$

Table 31: Base performance metrics for the trinary classifier.

Variable	Expression	Base Metrics
ρ_C	Proportion of inputs in the test set that belong to class C	$1 - \rho_A - \rho_B$
φ_C	$\dfrac{FC^A}{TA + FB^A + FC^A}$	$1 - \varphi_A - \varphi_B$
χ_C	$\dfrac{FC^B}{FA^B + TB + FC^B}$	$1 - \chi_A - \chi_B$
ψ_C	$\dfrac{TC}{FA^C + FB^C + TC}$	$1 - \psi_A - \psi_B$

Table 32: Auxiliary performance metrics.

Table 32 list the auxiliary metrics. These metrics can be written in terms of the base metrics from Table 31 as shown in the third column. However, we use these symbols for convenience when specifying equations in terms of the base metrics.

5.2-b CONSTRAINTS

From the previous section we see that each column in the confusion matrix must add to the proportion of test inputs for the class in the column. Each of these represents a constraint on the system.

The confusion matrix has nine elements. Each column has one constraint. Together, there are six free variables for the trinary classifier. The binary classifier had four parameters and two constraints, which left us with two variables.

We only needed two variables to characterize a binary classifier. The trinary classifier requires six variables to characterize the classifier. This also means we have six dimensions in the ROC space for the trinary classifier. Because of this, ROC plots are less common for trinary classifiers. However, the basic ROC analysis can still be used if extended to six variables.

Each column of the matrix in Figure 48 is independent of the other columns. The distributions of the variables are independent across columns. However, the variables in the same column are not independent of each other.

The tally of the numbers in the confusion matrix is distributed similar to the binomial distribution. However, in this case there are two distinct variables. Let x and y be the variables and let $z = 1 - x - y$. If we have N trials, the distribution takes the form,

The probability of receiving k x's and l y's is given by the multinomial distribution

$$\frac{N!}{k!\,l!\,(N - k - l)!} x^k y^l (1 - x - y)^{N-k-l} \qquad 5.8$$

Alternatively, we can examine this from the perspective of a distribution with fixed k and l, where the values x and y vary. The distribution of these variables is

$$B(k, l, N; x, y) = A x^k y^l (1 - x - y)^{N-k-l}$$
$$0 \leq x \leq 1 \quad 0 \leq y \leq 0 \quad x + y \leq 1 \qquad 5.9$$

where A is the normalization.

Examine the distribution of the form

$$B(k, l, N; x, y) = \frac{\Gamma(m + n + p)}{\Gamma(m)\Gamma(n)\Gamma(p)} x^{m-1} y^{n-1} (1 - x - y)^{p-1} \qquad 5.10$$

$$m, n, p > 0 \quad 0 \leq x \leq 1 \quad 0 \leq y \leq 0 \quad x + y \leq 1$$

This is the probability density of the bivariate Dirichlet distribution. The mean and variance of the distribution with respect to x and y are:

$$\mu_x = \frac{m}{m + n + p} \quad \sigma_x^2 = \frac{m(n + p)}{(m + n + p)^2 (m + n + p + 1)}$$

$$\mu_y = \frac{n}{m + n + p} \quad \sigma_y^2 = \frac{n(m + p)}{(m + n + p)^2 (m + n + p + 1)}$$

5.11

The multinomial distribution is

$$\frac{N!}{k! \, l! \, m!} x^k y^l (1 - x - y)^m$$

5.12

If we scale this to the range [0,1], the mean and standard are

$$\mu_x = k \quad \sigma_x^2 = \frac{k(1 - k)}{N}$$

$$\mu_y = l \quad \sigma_y^2 = \frac{l(1 - l)}{N}$$

5.13

We can use the method of moments to find expressions for m, n, and p in terms of k, l, and N from 5.8:

$$m = k(N - 1)$$
$$n = l(N - 1)$$
$$p = (N - 1)(1 - k - l)$$

5.14

In terms of the variables from the base matrix, we have three independent Dirichlet distributions where the parameters are

$$m = \varphi_A(\rho_A N - 1)$$
$$n = \varphi_B(\rho_A N - 1)$$
$$p = (\rho_A N - 1)(1 - \varphi_A - \varphi_B)$$

5.15

$$m = \chi_A(\rho_B N - 1)$$
$$n = \chi_B(\rho_B N - 1)$$
$$p = (\rho_B N - 1)(1 - \chi_A - \chi_B)$$

5.16

$$m = \psi_A(\rho_C N - 1)$$
$$n = \psi_B(\rho_C N - 1)$$
$$p = (\rho_C N - 1)(1 - \psi_A - \psi_B)$$

5.17

5.2-c TRINARY CLASSIFIER METRICS

In the previous section, we found that the trinary classifier requires six parameters to characterize the behavior. The base metrics for the trinary classifier are provided in Table 31. In this section, we develop related performance measures and specify them in terms of the base metrics.

The binary classifier metrics are provided in Table 18 in terms of the base metrics, and in Table 11 in terms of the elements of the confusion matrix. Looking at Table 11, we see that eight of the metrics (TPR, FRP, TNR, FNR, PPC, NPV, FDR, and NDR) are written in terms of probability measures, four are expressions for the confusion matrix, two are related to the breakdown of the test input set (AP and AN), two are related to the breakdown of the input set by the classifier (CP and CN), one is a measure of overall accuracy (ACC), and two are hybrid metrics designed to provide a single overall score for the classifier. We generalize all of these, except for the hybrid metrics, for the trinary classifier.

5.2-c(i) INPUT SET METRICS

The input set metrics for the binary classifier are AP and AN. These measure the actual number of positive and negative inputs in the test input set. For the trinary classifier, we have three such numbers, one counting the number of inputs in each of the three classes A, B, and C.

These metrics are generalized to the trinary classifier as:

$$A^A = TA + FB^A + FC^A \qquad 5.18$$

$$A^B = FA^B + TB + FC^B \qquad 5.19$$

$$A^C = FA^C + FB^C + TC \qquad 5.20$$

These three metrics are the sums of each of the columns in the confusion matrix. We can also write these in terms of the base metrics as

$$A^A = \rho_A N \qquad 5.21$$

$$A^B = \rho_B N \qquad 5.22$$

$$A^A = (1 - \rho_A - \rho_B)N \qquad 5.23$$

5.2-c(ii) CLASSIFIED SET METRICS

With the binary classifier, we used CP and CN as metrics for how the classifier placed inputs into categories. CP is the number of inputs the classifier placed into the positive class, while CN is the number of inputs the classifier placed into the negative class. Extending this,

$$C^A = TA + FA^B + FA^C \qquad 5.24$$

$$C^B = FB^A + TB + FB^C \qquad \text{5.25}$$

$$C^C = FC^A + FC^B + TC \qquad \text{5.26}$$

These metrics are the sums of each of the rows of the confusion matrix. In terms of the base metrics,

$$C^A = (\rho_A \varphi_A + \rho_B \chi_A + \rho_C \psi_A)N \qquad \text{5.27}$$

$$C^B = (\rho_A \varphi_B + \rho_B \chi_B + \rho_C \psi_B)N \qquad \text{5.28}$$

$$C^A = (\rho_A \varphi_C + \rho_B \chi_C + \rho_C \psi_C)N \qquad \text{5.29}$$

5.2-c(iii) CONFUSION MATRIX METRICS

The elements of the confusion matrix can also be written in terms of the base metrics. Examining Table 31 and Table 32, the denominators appearing in each of the metrics is the sum of the cells in the respective column. This is just one of the input set metrics. We can multiply through by this denominator to solve for each element of the confusion matrix in terms of the base metrics.

$$TA = \varphi_A \rho_A N \qquad \text{5.30}$$

$$FB^A = \varphi_B \rho_A N \qquad \text{5.31}$$

$$FC^A = \varphi_C \rho_A N \qquad \text{5.32}$$

$$FA^B = \chi_A \rho_B N \qquad \text{5.33}$$

$$TB = \chi_B \rho_B N \qquad \text{5.34}$$

$$FC^B = \chi_C \rho_B N \qquad \text{5.35}$$

$$FA^C = \psi_A \rho_C N \qquad \text{5.36}$$

$$FB^C = \psi_B \rho_C N \qquad \text{5.37}$$

$$TC = \psi_C \rho_C N \qquad \text{5.38}$$

5.2-c(iv) PROBABILITY METRICS

The probability metrics for the trinary classifier are more complicated than the binary classifier because there are multiple ways we can extend the probability concepts.

The binary classifier examined eight conditional probabilities such as $P(T_A|C_A)$. This is the probability that the true value of an input is A given the classifier determined that the input is A. This is easily extended to the trinary classifier.

However, probabilities such as $P(T_A|C_{\bar{A}})$ can be extended in two different ways. This is the probability that the true value of an input is A given that the classifier determined that the input is not A. The 'not A' portion can be extended as 'not A' meaning either B or C, or this portion can be extended to just B or just C. Thus, we could have $P(T_A|C_B \lor C_C)$, $P(T_A|C_B)$, or $P(T_A|C_C)$. Both of these extensions are interesting.

The extension $P(T_A|C_B \lor C_C)$ tells us the probability that the true value is A when the classifier puts the input into another category. This helps us understand how confident we are about the classifiers negative predictions. The second extension tells us about the correlative behaviors of the classifier. When the true value is A and the classifier is wrong, is the classifier more likely to put the input into B or C? Both of these are interesting results.

Finally, probabilities like $P(T_{\bar{A}}|C_{\bar{A}})$ can be extended in even more ways. Here we have two negations. Each could be extended in the same manner as above. For example, $P(T_{\bar{A}}|C_{\bar{A}})$ can be extended as: $P(T_B \lor T_C|C_B \lor C_C)$, $P(T_A|C_B \lor C_C)$, $P(T_B|C_B \lor C_C)$, $P(T_B \lor T_C|C_B)$, $P(T_B \lor T_C|C_C)$, $P(T_B|C_B)$, $P(T_B|C_C)$, $P(T_C|C_B)$, or $P(T_C|C_C)$.

In total, we have six possibilities for the true values events, and six events for the classified values. The true values events are: $T_A, T_B, T_C, T_A \lor T_B, T_A \lor T_C$, and $T_B \lor T_C$. Similarly, the classified values are: $C_A, C_B, C_C, C_A \lor C_B, C_A \lor C_C$, and $C_B \lor C_C$.

We can pair each true value with each classified value to create a metric. In addition, we can examine true value given classified value, or classified value given true value. In total there are 72 different metrics.

These probability metrics can be found from the confusion matrix. For example, $P(T_A|C_C)$ can be found by examining the A column and C row in the confusion matrix. The denominator is the sum of the cells of the row, while the numerator is the cell where the column and row intersect. Figure 49 shows the cells of the confusion matrix involved in computing this metric.

The probability $P(C_A|B_B)$ is computed similarly and shown in Figure 50. In this case, the denominator is the sum of the column, while the numerator is the cell where the column and row intersect.

Generally, the numerator is the row where the column and row intersect. The denominator is the sum of the row if the classified value is the given (event after the | in the expression) in the probability. Alternatively, if the true value is the given, then the sum over the cells in the column is the given.

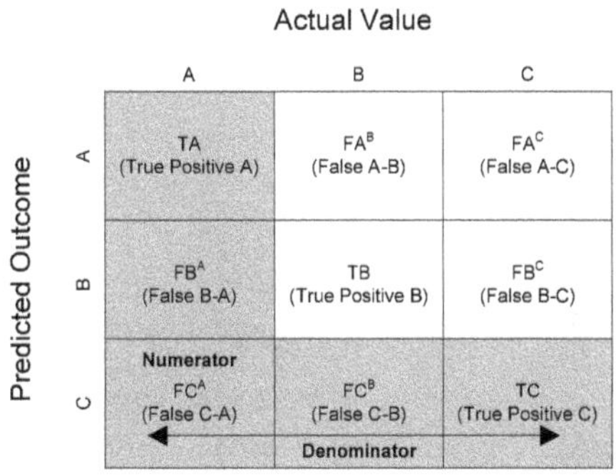

Figure 49: Computation of $P(T_A|C_C)$. Identify the column corresponding to T_A and the row corresponding to C_C. The denominator of $P(T_A|C_C)$ is the sum of the elements in the row. The numerator is given by the cell where the row and column intersect.

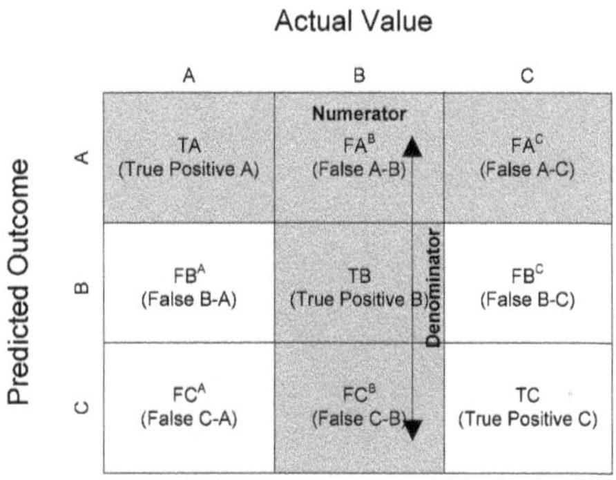

Figure 50: Computation of $P(C_A|T_B)$. Identify the column corresponding to C_A and the row corresponding to T_B. The denominator of $P(T_A|C_C)$ is the sum of the elements in the column. The numerator is given by the cell where the row and column intersect.

More complicated expressions such as $P(T_B \vee T_C|C_B)$ can be computed using similar reasoning. For this expression, the denominator remains unchanged (the sum of the cells for the B row). The numerator is the result of 'or-ing' the events B and C:

$$N = T_B \vee T_C = T_B + T_C - T_B \cap T_C \qquad 5.39$$

The expression $T_B \cap T_C$ means the number of events that are assigned to both classes B and C. Thus far we have assumed that each input is assigned to only one class. Classifiers that assign each input to exactly one class are pigeonhole classifiers. Combination classifiers allow inputs to be placed in multiple classes. For a pigeonhole classifier, $T_B \cap T_C = \emptyset$, so there are no such events and this term may be dropped.

We will proceed assuming $T_B \cap T_C = \emptyset$. A combination classifier can be made into a binary classifier by adding additional classes. When an input is placed into multiple classes, we simply treat this as a different class. If we make a separate class for each possible combination, then each input is placed into exactly one class. Thus, these metrics can be used for the combination classifier as well, so long as we add additional categories to handle cases where inputs are placed into multiple classes.

For the pigeonhole classifier, if we have an $\vee$ appearing for the numerator, we examine the row[14] (column) for each event appearing in the $\vee$, then identify the cells that intersect the column (row) for the denominator. The numerator is the sum of the intersecting cells.

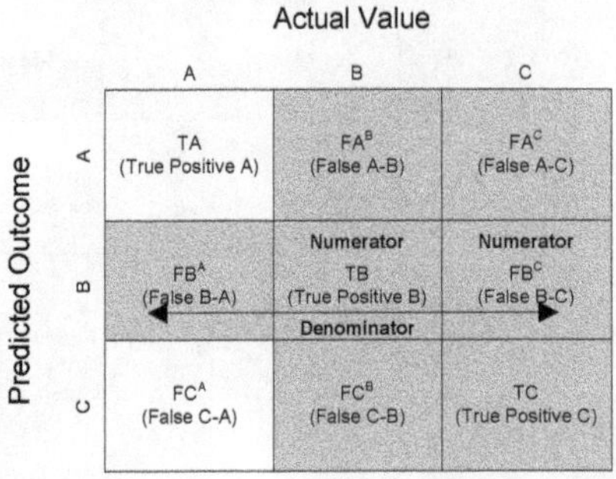

Figure 51: Computation of $P(T_B \vee T_C | C_B)$. Identify the columns corresponding to T_B and T_C, along with the row corresponding to C_B. The denominator is the sum of the elements in the row. The numerator is sum of the cells where the row and columns intersect.

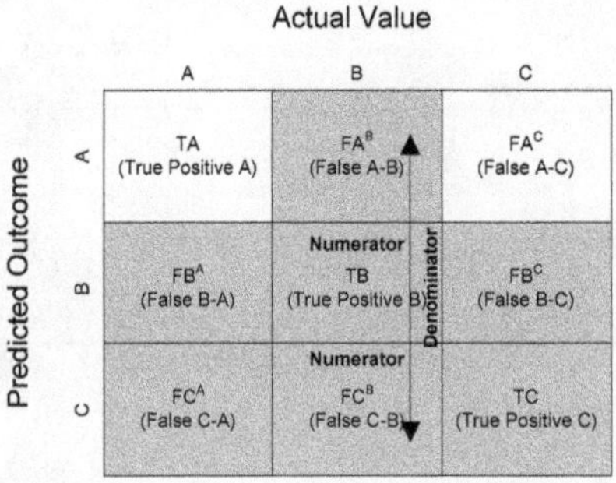

Figure 52: Computation of $P(C_B \vee C_C | T_B)$. Identify the rows corresponding to C_B and C_C, and the column corresponding to T_B. The denominator is the sum of the elements in the column. The numerator is the sum of the cells where the rows and column intersect.

The case for $P(C_B \vee C_C | T_B)$ is similar. Here, the denominator is the sum of the elements in the column for T_B. The numerator is the sum of the cells where the rows for C_B and C_C intersect the column for T_B.

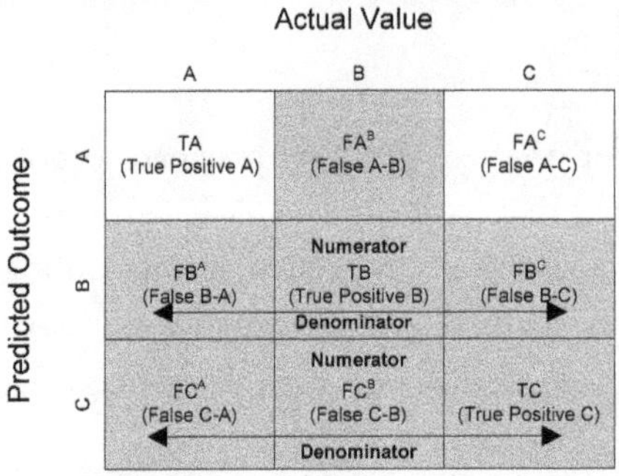

Figure 53: Computation of $P(T_B | C_B \vee C_C)$. Identify the rows corresponding to C_B and C_C, along with the column corresponding to T_B. The denominator is the sum of the elements in the two rows. The numerator is sum of the cells where the rows and column intersect.

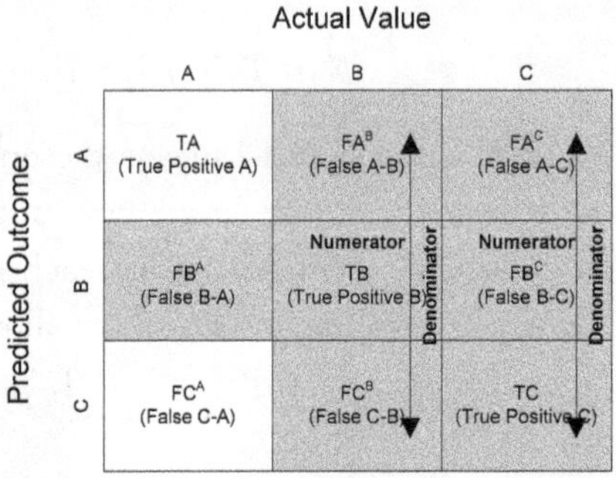

Figure 54: Computation of $P(C_B | T_B \vee T_C)$. Identify the columns corresponding to T_B and T_C, and the row corresponding to C_B. The denominator is the sum of the elements in the columns. The numerator is the sum of the cells where the row and columns intersect.

---

[14] We use the row if the numerator corresponds to a true value, and the column if we are examining the classified value.

Next, examine the case for $P(T_B|C_B \lor C_C)$. Here, the denominator is the sum of the cells over the rows for both C_B and C_C. The numerator is the sum of the cells where T_B intersects the rows from C_B and C_C. Similarly, $P(C_B|T_B \lor T_C)$ has a denominator which is the sum of the cells in the columns from T_B and T_C. The numerator is the sum of the cells where the row C_B intersects the columns T_C and T_C.

Finally we consider the case where there is a $\lor$ on both sides of the given ($|$). The probability $P(T_A \lor T_C|C_B \lor C_C)$ is such a case. Here, we identify the two columns corresponding to T_A and T_C, and the two rows corresponding to C_B and C_C. The denominator is the sum of all of the cells in the rows. The numerator is the sum of the four cells where the columns T_A and T_C intersect the rows C_B and C_C. This case is shown in Figure 55.

The case for $P(C_B \lor C_C|T_A \lor T_C)$ is handled with a similar method. Again, we identify the rows corresponding to C_B and C_C and the columns corresponding to T_A and T_C. In this case, the denominator is the sum of the cells in the columns, while the numerator is the sum of the cells where the rows intersect the columns.

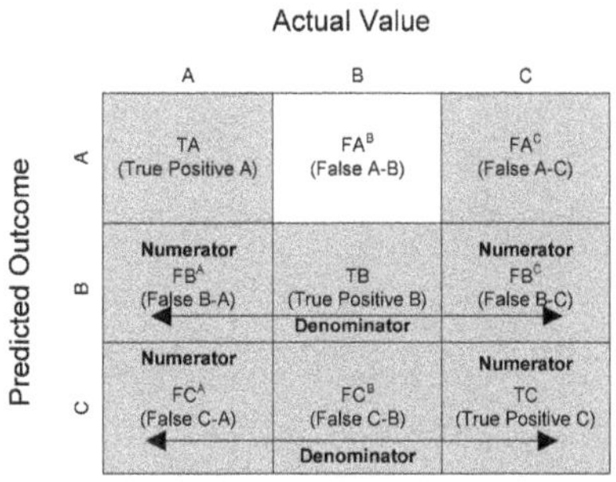

Figure 55: Computation of $P(T_A \lor T_C|C_B \lor C_C)$. Identify the rows corresponding to C_B and C_C, along with the columns corresponding to T_A and T_C. The denominator is the sum of the elements in the two rows. The numerator is sum of the cells where the rows and columns intersect.

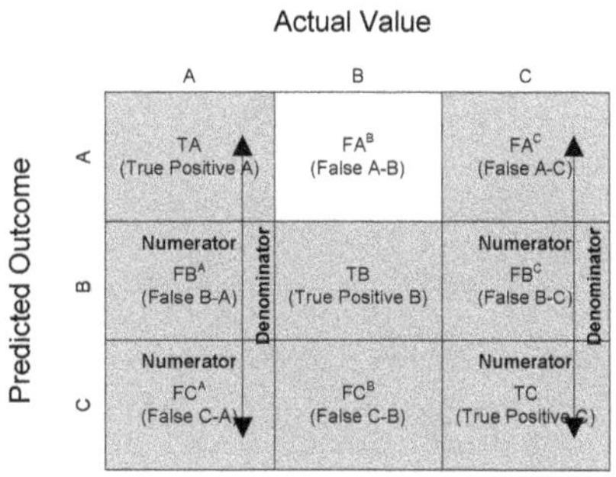

Figure 56: Computation of $P(C_B|T_B \lor T_C)$. Identify the columns corresponding to T_B and T_C, and the row corresponding to C_B. The denominator is the sum of the elements in the columns. The numerator is the sum of the cells where the row and columns intersect.

This covers the cases that arise for the probability metrics in the trinary classifier. We can use these methods to compute each of the 72 different probability metrics. We divide the cases into two sets depending on the form of the probability: $P(C|T)$ and $P(T|C)$.

$P(T\|C)$	T_A	T_B	T_C	$T_A \vee T_B$	$T_A \vee T_C$	$T_B \vee T_C$
C_A	$\dfrac{TA}{TA+FA^B+FA^C}$	$\dfrac{FA^B}{TA+FA^B+FA^C}$	$\dfrac{FA^C}{TA+FA^B+FA^C}$	$\dfrac{TA+FA^B}{TA+FA^B+FA^C}$	$\dfrac{TA+FA^C}{TA+FA^B+FA^C}$	$\dfrac{FA^B+FA^C}{TA+FA^B+FA^C}$
C_B	$\dfrac{FB^A}{FB^A+TB+FB^C}$	$\dfrac{TB}{FB^A+TB+FB^C}$	$\dfrac{FB^C}{FB^A+TB+FB^C}$	$\dfrac{FB^A+TB}{FB^A+TB+FB^C}$	$\dfrac{FB^A+FB^C}{FB^A+TB+FB^C}$	$\dfrac{TB+FB^C}{FB^A+TB+FB^C}$
C_C	$\dfrac{FC^A}{FC^A+FC^B+TC}$	$\dfrac{FC^B}{FC^A+FC^B+TC}$	$\dfrac{TC}{FC^A+FC^B+TC}$	$\dfrac{FC^A+FC^B}{FC^A+FC^B+TC}$	$\dfrac{FC^A+TC}{FC^A+FC^B+TC}$	$\dfrac{FC^B+TC}{FC^A+FC^B+TC}$

Table 33: Probability metrics for $P(T\|C)$ for the trinary classifier in terms of the elements of the confusion matrix.

$P(T\|C)$	T_A	T_B	T_C
$C_A \vee C_B$	$\dfrac{TA+FB^A}{TA+FA^B+FA^C+FB^A+TB+FB^C}$	$\dfrac{FA^B+TB}{TA+FA^B+FA^C+FB^A+TB+FB^C}$	$\dfrac{FA^C+FB^C}{TA+FA^B+FA^C+FB^A+TB+FB^C}$
$C_A \vee C_C$	$\dfrac{TA+FC^A}{TA+FA^B+FA^C+FC^A+FC^B+TC}$	$\dfrac{FA^B+FC^B}{TA+FA^B+FA^C+FC^A+FC^B+TC}$	$\dfrac{FA^C+TC}{TA+FA^B+FA^C+FC^A+FC^B+TC}$
$C_B \vee C_C$	$\dfrac{FB^A+FC^A}{FB^A+TB+FB^C+FC^A+FC^B+TC}$	$\dfrac{TB+FC^B}{FB^A+TB+FB^C+FC^A+FC^B+TC}$	$\dfrac{FB^C+TC}{FB^A+TB+FB^C+FC^A+FC^B+TC}$

Table 34: Probability metrics for $P(T\|C)$ for the trinary classifier in terms of the elements of the confusion matrix.

$P(T\|C)$	$T_A \vee T_B$	$T_A \vee T_C$	$T_B \vee T_C$
$C_A \vee C_B$	$\dfrac{TA+FB^A+FA^B+TB}{TA+FA^B+FA^C+FB^A+TB+FB^C}$	$\dfrac{TA+FB^A+FA^C+FB^C}{TA+FA^B+FA^C+FB^A+TB+FB^C}$	$\dfrac{FA^B+TB+FA^C+FB^C}{TA+FA^B+FA^C+FB^A+TB+FB^C}$
$C_A \vee C_C$	$\dfrac{TA+FC^A+FA^B+FC^B}{TA+FA^B+FA^C+FC^A+FC^B+TC}$	$\dfrac{TA+FC^A+FA^C+TC}{TA+FA^B+FA^C+FC^A+FC^B+TC}$	$\dfrac{FA^B+FC^B+FA^C+TC}{TA+FA^B+FA^C+FC^A+FC^B+TC}$
$C_B \vee C_C$	$\dfrac{FB^A+FC^A+TB+FC^B}{FB^A+TB+FB^C+FC^A+FC^B+TC}$	$\dfrac{FB^A+FC^A+FB^C+TC}{FB^A+TB+FB^C+FC^A+FC^B+TC}$	$\dfrac{TB+FC^B+FB^C+TC}{FB^A+TB+FB^C+FC^A+FC^B+TC}$

Table 35: Probability metrics for $P(T\|C)$ for the trinary classifier in terms of the elements of the confusion matrix.

$P(C\|T)$	C_A	C_B	C_C	$C_A \vee C_B$	$C_A \vee C_C$	$C_B \vee C_C$
T_A	$\dfrac{TA}{TA+FB^A+FC^A}$	$\dfrac{FB^A}{TA+FB^A+FC^A}$	$\dfrac{FC^A}{TA+FB^A+FC^A}$	$\dfrac{TA+FB^A}{TA+FB^A+FC^A}$	$\dfrac{TA+FC^A}{TA+FB^A+FC^A}$	$\dfrac{FB^A+FC^A}{TA+FB^A+FC^A}$
T_B	$\dfrac{FA^B}{FA^B+TB+FC^B}$	$\dfrac{TB}{FA^B+TB+FC^B}$	$\dfrac{FC^B}{FA^B+TB+FC^B}$	$\dfrac{FA^B+TB}{FA^B+TB+FC^B}$	$\dfrac{FA^B+FC^B}{FA^B+TB+FC^B}$	$\dfrac{TB+FC^B}{FA^B+TB+FC^B}$
T_C	$\dfrac{FA^C}{FA^C+FB^C+TC}$	$\dfrac{FB^C}{FA^C+FB^C+TC}$	$\dfrac{TC}{FA^C+FB^C+TC}$	$\dfrac{FA^C+FB^C}{FA^C+FB^C+TC}$	$\dfrac{FA^C+TC}{FA^C+FB^C+TC}$	$\dfrac{FB^C+TC}{FA^C+FB^C+TC}$

Table 36: Probability metrics for $P(C\|T)$ for the trinary classifier in terms of the elements of the confusion matrix.

$P(C\|T)$	C_A	C_B	C_C
$T_A \vee T_B$	$\dfrac{TA+FA^B}{TA+FB^A+FC^A+FA^B+TB+FC^B}$	$\dfrac{FB^A+TB}{TA+FB^A+FC^A+FA^B+TB+FC^B}$	$\dfrac{FC^A+FC^B}{TA+FB^A+FC^A+FA^B+TB+FC^B}$
$T_A \vee T_C$	$\dfrac{TA+FA^C}{TA+FB^A+FC^A+FA^C+FB^C+TC}$	$\dfrac{FB^A+FB^C}{TA+FB^A+FC^A+FA^C+FB^C+TC}$	$\dfrac{FC^A+TC}{TA+FB^A+FC^A+FA^C+FB^C+TC}$
$T_B \vee T_C$	$\dfrac{FA^B+FA^C}{FA^B+TB+FC^B+FA^C+FB^C+TC}$	$\dfrac{TB+FB^C}{FA^B+TB+FC^B+FA^C+FB^C+TC}$	$\dfrac{FC^B+TC}{FA^B+TB+FC^B+FA^C+FB^C+TC}$

Table 37: Probability metrics for $P(C\|T)$ for the trinary classifier in terms of the elements of the confusion matrix.

$P(C\|T)$	$C_A \vee C_B$	$C_A \vee C_C$	$C_B \vee C_C$
$T_A \vee T_B$	$\dfrac{TA + FA^B + FB^A + TB}{TA + FB^A + FC^A + FA^B + TB + FC^B}$	$\dfrac{TA + FA^B + FC^A + FC^B}{TA + FB^A + FC^A + FA^B + TB + FC^B}$	$\dfrac{FB^A + TB + FC^A + FC^B}{TA + FB^A + FC^A + FA^B + TB + FC^B}$
$T_A \vee T_C$	$\dfrac{TA + FA^C + FB^A + FB^C}{TA + FB^A + FC^A + FA^C + FB^C + TC}$	$\dfrac{TA + FA^C + FC^A + TC}{TA + FB^A + FC^A + FA^C + FB^C + TC}$	$\dfrac{FB^A + FB^C + FC^A + TC}{TA + FB^A + FC^A + FA^C + FB^C + TC}$
$T_B \vee T_C$	$\dfrac{FA^B + FA^C + TB + FB^C}{FA^B + TB + FC^B + FA^C + FB^C + TC}$	$\dfrac{FA^B + FA^C + FC^B + TC}{FA^B + TB + FC^B + FA^C + FB^C + TC}$	$\dfrac{TB + FB^C + FC^B + TC}{FA^B + TB + FC^B + FA^C + FB^C + TC}$

Table 38: Probability metrics for $P(C|T)$ for the trinary classifier in terms of the elements of the confusion matrix.

$P(T\|C)$	T_A	T_B	T_C	$T_A \vee T_B$	$T_A \vee T_C$	$T_B \vee T_C$
C_A	$\dfrac{\varphi_A \rho_A}{\varphi_A \rho_A + \chi_A \rho_B + \psi_A \rho_C}$	$\dfrac{\chi_A \rho_B}{\varphi_A \rho_A + \chi_A \rho_B + \psi_A \rho_C}$	$\dfrac{\psi_A \rho_C}{\varphi_A \rho_A + \chi_A \rho_B + \psi_A \rho_C}$	$\dfrac{\varphi_A \rho_A + \chi_A \rho_B}{\varphi_A \rho_A + \chi_A \rho_B + \psi_A \rho_C}$	$\dfrac{\varphi_A \rho_A + \psi_A \rho_C}{\varphi_A \rho_A + \chi_A \rho_B + \psi_A \rho_C}$	$\dfrac{\chi_A \rho_B + \psi_A \rho_C}{\varphi_A \rho_A + \chi_A \rho_B + \psi_A \rho_C}$
C_B	$\dfrac{\varphi_B \rho_A}{\varphi_B \rho_A + \chi_B \rho_B + \psi_B \rho_C}$	$\dfrac{\chi_B \rho_B}{\varphi_B \rho_A + \chi_B \rho_B + \psi_B \rho_C}$	$\dfrac{\psi_B \rho_C}{\varphi_B \rho_A + \chi_B \rho_B + \psi_B \rho_C}$	$\dfrac{\varphi_B \rho_A + \chi_B \rho_B}{\varphi_B \rho_A + \chi_B \rho_B + \psi_B \rho_C}$	$\dfrac{\varphi_B \rho_A + \psi_B \rho_C}{\varphi_B \rho_A + \chi_B \rho_B + \psi_B \rho_C}$	$\dfrac{\chi_B \rho_B + \psi_B \rho_C}{\varphi_B \rho_A + \chi_B \rho_B + \psi_B \rho_C}$
C_C	$\dfrac{\varphi_C \rho_A}{\varphi_C \rho_A + \chi_C \rho_B + \psi_C \rho_C}$	$\dfrac{\chi_C \rho_B}{\varphi_C \rho_A + \chi_C \rho_B + \psi_C \rho_C}$	$\dfrac{\psi_C \rho_C}{\varphi_C \rho_A + \chi_C \rho_B + \psi_C \rho_C}$	$\dfrac{\varphi_C \rho_A + \chi_C \rho_B}{\varphi_C \rho_A + \chi_C \rho_B + \psi_C \rho_C}$	$\dfrac{\varphi_C \rho_A + \psi_C \rho_C}{\varphi_C \rho_A + \chi_C \rho_B + \psi_C \rho_C}$	$\dfrac{\chi_C \rho_B + \psi_C \rho_C}{\varphi_C \rho_A + \chi_C \rho_B + \psi_C \rho_C}$

Table 39: Probability metrics for $P(T|C)$ for the trinary classifier in terms of the elements of the base metrics.

$P(T\|C)$	T_A	T_B	T_C
$C_A \vee C_B$	$\dfrac{\varphi_A \rho_A + \varphi_B \rho_A}{\varphi_A \rho_A + \chi_A \rho_B + \psi_A \rho_C + \varphi_B \rho_A + \chi_B \rho_B + \psi_B \rho_C}$	$\dfrac{\chi_A \rho_B + \chi_B \rho_B}{\varphi_A \rho_A + \chi_A \rho_B + \psi_A \rho_C + \varphi_B \rho_A + \chi_B \rho_B + \psi_B \rho_C}$	$\dfrac{\psi_A \rho_C + \psi_B \rho_C}{\varphi_A \rho_A + \chi_A \rho_B + \psi_A \rho_C + \varphi_B \rho_A + \chi_B \rho_B + \psi_B \rho_C}$
$C_A \vee C_C$	$\dfrac{\varphi_A \rho_A + \varphi_C \rho_A}{\varphi_A \rho_A + \chi_A \rho_B + \psi_A \rho_C + \varphi_C \rho_A + \chi_C \rho_B + \psi_C \rho_C}$	$\dfrac{\chi_A \rho_B + \chi_C \rho_B}{\varphi_A \rho_A + \chi_A \rho_B + \psi_A \rho_C + \varphi_C \rho_A + \chi_C \rho_B + \psi_C \rho_C}$	$\dfrac{\psi_A \rho_C + \psi_C \rho_C}{\varphi_A \rho_A + \chi_A \rho_B + \psi_A \rho_C + \varphi_C \rho_A + \chi_C \rho_B + \psi_C \rho_C}$
$C_B \vee C_C$	$\dfrac{\varphi_B \rho_A + \varphi_C \rho_A}{\varphi_B \rho_A + \chi_B \rho_B + \psi_B \rho_C + \varphi_C \rho_A + \chi_C \rho_B + \psi_C \rho_C}$	$\dfrac{\chi_B \rho_B + \chi_C \rho_B}{\varphi_B \rho_A + \chi_B \rho_B + \psi_B \rho_C + \varphi_C \rho_A + \chi_C \rho_B + \psi_C \rho_C}$	$\dfrac{\psi_B \rho_C + \psi_C \rho_C}{\varphi_B \rho_A + \chi_B \rho_B + \psi_B \rho_C + \varphi_C \rho_A + \chi_C \rho_B + \psi_C \rho_C}$

Table 40: Probability metrics for $P(T|C)$ for the trinary classifier in terms of the elements of the base metrics

$P(T\|C)$	$T_A \vee T_B$	$T_A \vee T_C$	$T_B \vee T_C$
$C_A \vee C_B$	$\dfrac{\varphi_A \rho_A + \varphi_B \rho_A + \chi_A \rho_B + \chi_B \rho_B}{\varphi_A \rho_A + \chi_A \rho_B + \psi_A \rho_C + \varphi_B \rho_A + \chi_B \rho_B + \psi_B \rho_C}$	$\dfrac{\varphi_A \rho_A + \varphi_B \rho_A + \psi_A \rho_C + \psi_B \rho_C}{\varphi_A \rho_A + \chi_A \rho_B + \psi_A \rho_C + \varphi_B \rho_A + \chi_B \rho_B + \psi_B \rho_C}$	$\dfrac{\chi_A \rho_B + \chi_B \rho_B + \psi_A \rho_C + \psi_B \rho_C}{\varphi_A \rho_A + \chi_A \rho_B + \psi_A \rho_C + \varphi_B \rho_A + \chi_B \rho_B + \psi_B \rho_C}$
$C_A \vee C_C$	$\dfrac{\varphi_A \rho_A + \varphi_C \rho_A + \chi_A \rho_B + \chi_C \rho_B}{\varphi_A \rho_A + \chi_A \rho_B + \psi_A \rho_C + \varphi_C \rho_A + \chi_C \rho_B + \psi_C \rho_C}$	$\dfrac{\varphi_A \rho_A + \varphi_C \rho_A + \psi_A \rho_C + \psi_C \rho_C}{\varphi_A \rho_A + \chi_A \rho_B + \psi_A \rho_C + \varphi_C \rho_A + \chi_C \rho_B + \psi_C \rho_C}$	$\dfrac{\chi_A \rho_B + \chi_C \rho_B + \psi_A \rho_C + \psi_C \rho_C}{\varphi_A \rho_A + \chi_A \rho_B + \psi_A \rho_C + \varphi_C \rho_A + \chi_C \rho_B + \psi_C \rho_C}$
$C_B \vee C_C$	$\dfrac{\varphi_B \rho_A + \varphi_C \rho_A + \chi_B \rho_B + \chi_C \rho_B}{\varphi_B \rho_A + \chi_B \rho_B + \psi_B \rho_C + \varphi_C \rho_A + \chi_C \rho_B + \psi_C \rho_C}$	$\dfrac{\varphi_B \rho_A + \varphi_C \rho_A + \psi_B \rho_C + \psi_C \rho_C}{\varphi_B \rho_A + \chi_B \rho_B + \psi_B \rho_C + \varphi_C \rho_A + \chi_C \rho_B + \psi_C \rho_C}$	$\dfrac{\chi_B \rho_B + \chi_C \rho_B + \psi_B \rho_C + \psi_C \rho_C}{\varphi_B \rho_A + \chi_B \rho_B + \psi_B \rho_C + \varphi_C \rho_A + \chi_C \rho_B + \psi_C \rho_C}$

Table 41: Probability metrics for $P(T|C)$ for the trinary classifier in terms of the elements of the base metrics.

$P(C|T)$	C_A	C_B	C_C	$C_A \vee C_B$	$C_A \vee C_C$	$C_B \vee C_C$
T_A	φ_A	φ_B	φ_C	$\varphi_A + \varphi_B$	$\varphi_A + \varphi_C$	$\varphi_B + \varphi_C$
T_B	χ_A	χ_B	χ_C	$\chi_A + \chi_B$	$\chi_A + \chi_C$	$\chi_B + \chi_C$
T_C	ψ_A	ψ_B	ψ_C	$\psi_A + \psi_B$	$\psi_A + \psi_C$	$\psi_B + \psi_C$

Table 42: Probability metrics for $P(C|T)$ for the trinary classifier in terms of the elements of the base metrics.

$P(C|T)$	C_A	C_B	C_C
$T_A \vee T_B$	$\dfrac{\varphi_A \rho_A + \chi_A \rho_B}{\rho_A + \rho_B}$	$\dfrac{\varphi_B \rho_A + \chi_B \rho_B}{\rho_A + \rho_B}$	$\dfrac{\varphi_C \rho_A + \chi_C \rho_B}{\rho_A + \rho_B}$
$T_A \vee T_C$	$\dfrac{\varphi_A \rho_A + \psi_A \rho_C}{\rho_A + \rho_C}$	$\dfrac{\varphi_B \rho_A + \psi_B \rho_C}{\rho_A + \rho_C}$	$\dfrac{\varphi_C \rho_A + \psi_C \rho_C}{\rho_A + \rho_C}$
$T_B \vee T_C$	$\dfrac{\chi_A \rho_B + \psi_A \rho_C}{\rho_B + \rho_C}$	$\dfrac{\chi_B \rho_B + \psi_B \rho_C}{\rho_B + \rho_C}$	$\dfrac{\chi_C \rho_B + \psi_C \rho_C}{\rho_B + \rho_C}$

Table 43: Probability metrics for $P(C|T)$ for the trinary classifier in terms of the elements of the base metrics.

$P(C|T)$	$C_A \vee C_B$	$C_A \vee C_C$	$C_B \vee C_C$
$T_A \vee T_B$	$\dfrac{(\varphi_A + \varphi_B)\rho_A + (\chi_A + \chi_B)\rho_B}{\rho_A + \rho_B}$	$\dfrac{(\varphi_A + \varphi_C)\rho_A + (\chi_A + \chi_C)\rho_B}{\rho_A + \rho_B}$	$\dfrac{(\varphi_B + \varphi_C)\rho_A + (\chi_B + \chi_C)\rho_B}{\rho_A + \rho_B}$
$T_A \vee T_C$	$\dfrac{(\varphi_A + \varphi_B)\rho_A + (\psi_A + \psi_B)\rho_C}{\rho_A + \rho_C}$	$\dfrac{(\varphi_A + \varphi_C)\rho_A + (\psi_A + \psi_C)\rho_C}{\rho_A + \rho_C}$	$\dfrac{(\varphi_B + \varphi_C)\rho_A + (\psi_B + \psi_C)\rho_C}{\rho_A + \rho_C}$
$T_B \vee T_C$	$\dfrac{(\chi_A + \chi_B)\rho_B + (\psi_A + \psi_B)\rho_C}{\rho_B + \rho_C}$	$\dfrac{(\chi_A + \chi_C)\rho_B + (\psi_A + \psi_C)\rho_C}{\rho_B + \rho_C}$	$\dfrac{(\chi_B + \chi_C)\rho_B + (\psi_B + \psi_C)\rho_C}{\rho_B + \rho_C}$

Table 44: Probability metrics for $P(C|T)$ for the trinary classifier in terms of the elements of the base metrics.

Table 33-Table 44 provide the probability metrics for the trinary classifier. Table 33-Table 35 show the probability metrics for $P(T|C)$ in terms of the elements of the confusion matrix, while Table 36-Table 38 show the metrics for $P(C|T)$. The probability metrics can also be written in terms of the base metrics. Table 39-Table 41 provide the metrics for $P(T|C)$ in terms of the base metrics, while Table 42-Table 44 provide the metrics for $P(C|T)$ in terms of the base metrics.

5.2-c(v) ACCURACY

Accuracy is the ratio of the number of times the classifier is correct to the total number of inputs. For the binary classifier this is

$$ACC = \frac{TN + TP}{N} \qquad 5.40$$

The cells of the confusion matrix are show in Figure 57.

For the trinary classifier, the Accuracy is the sum of the diagonal elements of the confusion matrix, divided by the total number of inputs:

$$ACC = \frac{TA + TB + TC}{N} \qquad 5.41$$

These cells of the confusion matrix are show in Figure 58.

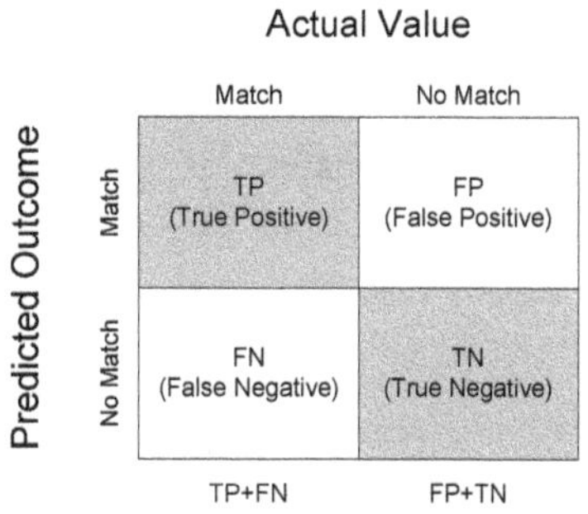

Figure 57: Confusion matrix elements used in the computation of Accuracy for the binary classifier.

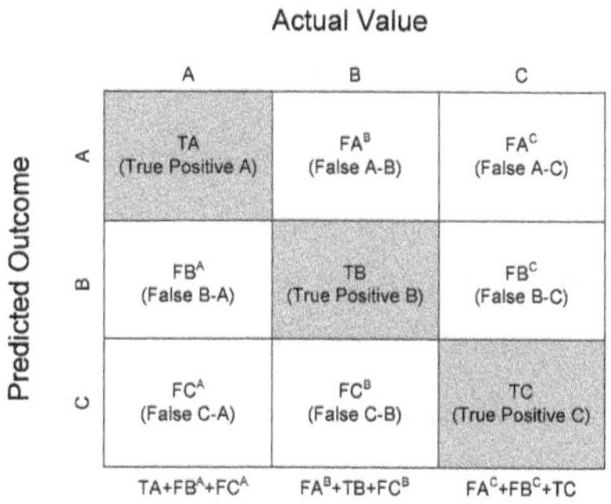

Figure 58: Confusion matrix elements used in the computation of Accuracy in the trinary classifier.

In terms of the base metrics, the trinary classifier metrics are

$$ACC = \varphi_A \rho_A + \chi_B \rho_B + \psi_C \rho_C \qquad 5.42$$

5.2-c(vi) DETERMINANT

The accuracy is written in terms of the trace of the confusion matrix divided by the total number of inputs. We can also examine the determinant of the confusion matrix. For the binary classifier this is

$$D = \frac{1}{N^2} \begin{vmatrix} TP & FP \\ FN & TN \end{vmatrix} \qquad 5.43$$

$$= \frac{TP \times TN - FP \times FN}{N^2} \qquad 5.44$$

In terms of the base metrics

$$= (\varphi - \chi)\rho(1 - \rho) \qquad 5.45$$

For the trinary classifier:

$$D = \frac{1}{N^3} \begin{vmatrix} TA & FA^B & FA^C \\ FB^A & TB & FB^C \\ FC^A & FC^B & TC \end{vmatrix} \qquad 5.46$$

In terms of the base metrics,

$$D = \begin{vmatrix} \varphi_A \rho_A & \chi_A \rho_B & \psi_A \rho_C \\ \varphi_B \rho_A & \chi_B \rho_B & \psi_B \rho_C \\ (1 - \varphi_A - \varphi_B)\rho_A & (1 - \chi_A - \chi_B)\rho_B & (1 - \psi_A - \psi_B)\rho_C \end{vmatrix} \qquad 5.47$$

$$D = [\varphi_A(\chi_B - \psi_B) - \chi_A(\varphi_B - \psi_B) + \psi_A(\varphi_B - \chi_B)]\rho_A \rho_B \rho_C \qquad 5.48$$

5.3 Error Analysis

A frequent question is how to compare the performance of two trinary classifiers. We assume that we have the confusion matrix and base metrics for each. How can we tell when the performance of one classifier is significantly (in the statistical sense) better than the other?

First, we need to identify what question we are really asking. The trinary classifier has three classes and six independent parameters. Each column of the confusion matrix has three parameters (not independent). One of the parameters indicates good performance (the diagonal elements) in the sense that the higher the value the better the classifier performs. The other two parameters both reflect bad performance (in the sense that the higher these values the worse the classifier performs).

The two parameters in each row tied to bad performance provide slightly different insight. When the classifier is wrong, it can be wrong in two ways, by incorrectly placing the input into one of two different categories. These parameters tells us which category we incorrectly placed the input.

For diagnostic purposes, the difference in these parameters may be of interest. For example, in a language classifier, did we mistake English for French or Italian? Knowing this helps direct our attention to make improvements to the classifier.

For comparison purposes, we typically are not interested in the nature of the mistake, only that the mistake is present. In some cases we may prefer mistakes to happen in a particular direction, but often this is not a concern.

We proceed with the assumption that we are only interested in the performance of the good parameter, and the bad parameters may be treated together. We begin by transforming the confusion matrix to just two rows: the first row is the count of the number of times the classifier is right for a given category, and the second is the number of times the classifier is wrong. This matrix may be written in terms of the confusion matrix or the base metrics. When we write this in terms of the base metrics, we call this the Positive Performance Matrix. Table 45 shows the performance matrix in terms of the element from the confusion matrix, while Table 46 shows the positive performance matrix.

	A	B	C
Correct	TA	TB	TC
Incorrect	$FB^A + FC^A$	$FA^B + FC^B$	$FB^C + FB^C$

Table 45: Performance matrix in terms of elements of the confusion matrix.

	A	B	C
Correct	φ_A	χ_B	ψ_C
Incorrect	$1 - \varphi_A$	$1 - \chi_B$	$1 - \psi_C$

Table 46: Positive Performance Matrix in terms of elements of the base matrix.

Generally, we write the positive performance matrix as

$$PPM = \begin{pmatrix} \varphi & \chi & \psi \\ 1 - \varphi & 1 - \chi & 1 - \psi \end{pmatrix} \qquad 5.49$$

where we set

$$\begin{aligned} \varphi &= \varphi_A \\ \chi &= \chi_B \\ \psi &= \psi_C \end{aligned} \qquad 5.50$$

The parameters in the positive performance matrix are independent of each other. In addition, these parameters each obey the Beta distribution rather than the Dirichlet distribution. The variance for these parameters is

$$\begin{aligned} \sigma_\varphi^2 &= \frac{\varphi(1 - \varphi)}{\rho_\varphi N} \\ \sigma_\chi^2 &= \frac{\chi(1 - \chi)}{\rho_\chi N} \\ \sigma_\psi^2 &= \frac{\psi(1 - \psi)}{\rho_\psi N} \end{aligned} \qquad 5.51$$

where N is the total number of inputs and $\rho_\varphi = \rho_A, \rho_\chi = \rho_B$, and $\rho_\psi = \rho_C$.

Given two trinary classifiers, each with the set of parameters from 5.50 with variances from 5.51, we wish to determine of one classifier is significantly 'better' than the other.

Let the subscript 1 indicate the first classifier and a subscript 2 indicate the second classifier. Given that the metrics are all independent and Beta distributed, we examine

$$P(\tilde{x}_1 \leq \tilde{x}_2 | x_1 > x_2) \qquad \text{5.52}$$

where $\tilde{x}$ indicates a random variable generated according to the distribution for x. This probability asks the question, given that we have measured $x_1 > x_2$, and given the distributions for x_1 and x_2, what is the probability that random variables could be generated where $\tilde{x}_1 \leq \tilde{x}_2$?

5.3-a MONTE CARLO

It may be difficult to obtain expressions for probabilities from the Beta distribution. Moreover, even numerical integration with the Beta distribution is difficult when the values for the parameters is large because we need to raise a number on the range [0,1] to a large power.

Instead of computing these directly from the Beta distribution, we can use a Monte Carlo approach to determining the probability. In this approach, we repeatedly generate a random Beta distributed random variable from each of the distributions for x_1 and x_2. Count the number of times $\tilde{x}_1 \leq \tilde{x}_2$, and divide this by the total number of random variable pairs we generate. This converges to the probability from 5.78.

Some computer software packages have built-in capabilities for generating Beta distributed random numbers. However, most at least have the ability to generate uniformly distributed random numbers over a predetermined range. From this, we can create a random number on the range [0,1]. This is a uniform deviate $U(0,1)$.

With a uniform deviate, we can create a Beta distributed random number as

$$B(\alpha, \beta) = \frac{\Gamma(1, \alpha)}{\Gamma(1, \alpha) + \Gamma(1, \beta)} \qquad \text{5.53}$$

where

$$\Gamma(1, \alpha) = -\ln \prod_{i=1}^{\alpha} U_i(0,1) \qquad \text{5.54}$$

$$\Gamma(1, \beta) = -\ln \prod_{i=1}^{\beta} U_i(0,1) \qquad \text{5.55}$$

This method is useful when both α abd β are integers.

5.3-b SIMULATION

Alternatively, we can simulate the entire classification process. We know that the number of test inputs that belong to the class is N_C, and the probability to successful classification of an input is x given that the input is actually in the class. We generate N_C uniform deviates, and count the number that are less than x. Do this for both the distribution for x_1 and x_2. If $\tilde{x}_1 \leq \tilde{x}_2$, add one to the count. As we repeat this process the count divided by the number of attempts approaches 5.78.

Either of these approaches may be used to estimate the probability from 5.78. We obtain one such probability by comparing each of the parameters φ, χ, and ψ. By examining each of these probabilities we can evaluate the significance of the relative performance of the classifiers.

5.3-c DECOMPOSITION TO BINARY CLASSIFIERS

Another approach is to treat the trinary classifier as three separate binary classifiers. Figure 59 indicates how the binary classifiers are constructed. For example, to construct the confusion matrix for the binary classifier resulting from class A, we set the TP value to the element on the diagonal for the row / column corresponding to class A. The FN value is the sum of the other elements in the same column, while the FP value is the sum of the elements in the same row. The TN value is the sum of the remaining elements.

The rationale behind this is that the true positive (TP) value for the binary classifier for A is the number of times the trinary classifier reported A when in fact the input presented was A. This is the TA cell of the confusion matrix.

The false negatives (FN) are the cells where the trinary classifier said something other than A when in fact the input was A. These are the cells in column A other than the TA cell.

The false positives (FP) are the cases where the classifier determined an input was A when in fact the input was something else. These are the cells in the same row as TA, other than the TA cell itself.

Finally, the true negatives (TN) are the cases where the classifier determined an input was not A when the input was actually not A. these are all of the remaining cells in the confusion matrix.

We compute these metrics for each class of the trinary classifier. This produces three independent performance metrics, corresponding to three binary classifiers. The performance metrics are independent because each set of metrics is drawn from a different set of inputs. So long as the input selection is uncorrelated, these metrics are uncorrelated as well.

		Actual Value		
		A	B	C
Predicted Outcome	A	TA (True Positive A)	FA^B (False A-B)	FA^C (False A-C)
	B	FB^A (False B-A)	TB (True Positive B)	FB^C (False B-C)
	C	FC^A (False C-A)	FC^B (False C-B)	TC (True Positive C)

		Actual Value		
		A	B	C
Predicted Outcome	A	TA (True Positive A)	FA^B (False A-B)	FA^C (False A-C)
	B	FB^A (False B-A)	TB (True Positive B)	FB^C (False B-C)
	C	FC^A (False C-A)	FC^B (False C-B)	TC (True Positive C)

		Actual Value		
		A	B	C
Predicted Outcome	A	TA (True Positive A)	FA^B (False A-B)	FA^C (False A-C)
	B	FB^A (False B-A)	TB (True Positive B)	FB^C (False B-C)
	C	FC^A (False C-A)	FC^B (False C-B)	TC (True Positive C)

Figure 59: Trinary classifier as three binary classifiers. Each class of the trinary classifier results in a separate binary classifier. The TP value is the element with both horizontal and vertical lines, the FN value is the sum of the cells with only vertical lines, the FP value is the sum of the calls with only horizontal lines, and the TN value is the sum of the remaining cells.

We can repeat this process for classes B and C as well. The result is three separate binary classifiers with the confusion matrix

$$A \qquad \begin{pmatrix} TA & FA^B + FA^C \\ FB^A + FC^A & TB + FB^C + FC^B + TC \end{pmatrix} \qquad 5.56$$

$$B \qquad \begin{pmatrix} TB & FB^A + FB^C \\ FA^B + FC^B & TA + FA^C + FC^A + TC \end{pmatrix} \qquad 5.57$$

$$C \qquad \begin{pmatrix} TC & FC^A + FC^B \\ FA^C + FB^C & TA + FA^B + FB^A + TB \end{pmatrix} \qquad 5.58$$

The base matrices are

$$A \qquad \begin{pmatrix} \varphi_A & \dfrac{\rho_B \chi_A + \rho_C \psi_A}{\rho_B + \rho_C} \\[2ex] 1 - \varphi_A & \dfrac{(1 - \chi_A)\rho_B + (1 - \psi_A)\rho_C}{\rho_B + \rho_C} \end{pmatrix} \qquad 5.59$$

$$B \qquad \begin{pmatrix} \chi_B & \dfrac{\rho_A \varphi_B + \rho_C \psi_B}{\rho_A + \rho_C} \\[2ex] 1 - \chi_B & \dfrac{(1 - \varphi_B)\rho_A + (1 - \psi_B)\rho_C}{\rho_A + \rho_C} \end{pmatrix} \qquad 5.60$$

$$C \qquad \begin{pmatrix} \psi_C & \dfrac{\rho_A \varphi_C + \rho_B \chi_C}{\rho_A + \rho_B} \\[2ex] 1 - \psi_C & \dfrac{(1 - \varphi_C)\rho_A + (1 - \chi_C)\rho_B}{\rho_A + \rho_B} \end{pmatrix} \qquad 5.61$$

We can examine each of these classifiers using the techniques developed previously in chapter 2. We can compare two such classifiers when the φ, χ, ψ from one classifier are all greater than the corresponding values from the second classifier. However, if some of these values are greater while others are smaller, the classifiers performance metrics are incomparable. We may still examine the relative performance, but we cannot simply judge one classifier as better than the other no matter how significant the statistics.

When some metrics are better while others are worse, we have the situation where one trinary classifier outperforms the second with respect to one category, while the second outperforms the first with respect to another category. This situation becomes increasingly common with multiclass classifiers as the total number of categories increases. In these cases, to determine a 'best' classifier, we need to develop some tradeoff function between the respective classes in order to determine the superior classifier.

Once we decompose the trinary classifier into three binary classifiers, we arrive at three independent performance metrics. We can extend each of the statistics from section 2.7 to the case of three metrics.

The distance statistic from 2.7-a is straightforward to extend to three metrics:

$$\frac{(\varphi_A - \bar{\varphi}_A)^2 + (\chi_B - \bar{\chi}_B)^2 + (\psi_C - \bar{\psi}_C)^2}{\sqrt{(\varphi_A - \bar{\varphi}_A)^2(\sigma_{\varphi_A}^2 + \sigma_{\bar{\varphi}_A}^2) + (\chi_B - \bar{\chi})^2(\sigma_{\chi_B}^2 + \sigma_{\bar{\chi}_B}^2) + (\psi_C - \bar{\psi}_C)^2\left(\sigma_{\psi_C}^2 + \sigma_{\bar{\psi}_C}^2\right)}}$$ 5.62

However, as the number of metrics increases, it becomes increasingly likely that one of the metrics will have a much smaller variance than the others. Since this statistic does not compute relative weights for each of the metrics, large differences in one metric can overpower smaller differences in the other metrics.

The statistic from 2.7-b can be extended to three variables as well. The distribution of the absolute value of a z-score is

$$K_1(z) = \begin{cases} \dfrac{2}{\sqrt{2\pi}}e^{-\frac{z^2}{2}} & z \geq 0 \\ 0 & z < 0 \end{cases}$$ 5.63

For the sum of the absolute value of two z-scores, the distribution is

$$K_2(z) = \begin{cases} \dfrac{2}{\sqrt{\pi}}e^{-\frac{z^2}{4}} \, erf(z/2) & z \geq 0 \\ 0 & z < 0 \end{cases}$$ 5.64

The sum of the absolute value of three z-scores is the convolution of these distributions

$$K_3(z) = \int_{-\infty}^{\infty} K_2(w)K_1(z-w)\,dw$$ 5.65

The integrand is zero when $w < 0$ or $w > z$. Thus,

$$K_3(z) = \int_0^z K_2(w) K_1(z - w) \, dw \qquad \text{5.66}$$

$$= \frac{4}{\sqrt{2\pi}} \int_0^z e^{-\frac{w^2}{4}} erf(w/2) e^{-\frac{(z-w)^2}{2}} \, dw \qquad \text{5.67}$$

$$= \frac{4}{\sqrt{2\pi}} e^{-\frac{z^2}{6}} \int_0^z e^{-\frac{3}{4}\left(w - \frac{2}{3}z\right)^2} erf(w/2) \, dw \qquad \text{5.68}$$

This integral can be computed numerically if desired. The confidence is the integral of this expression. Computing the confidence requires two successive numerical integrations. This complexity limits the utility of this statistics for trinary classifiers.

Hotelling's statistic from 2.7-c has a wider applicability. The statistic is

$$H_3 = \frac{(\varphi_A - \bar{\varphi}_A)^2}{\sigma_{\varphi_A}^2 + \sigma_{\bar{\varphi}_A}^2} + \frac{(\chi_B - \bar{\chi}_B)^2}{\sigma_{\chi_B}^2 + \sigma_{\bar{\chi}_B}^2} + \frac{(\psi_C - \bar{\psi}_C)^2}{\sigma_{\psi_C}^2 + \sigma_{\bar{\psi}_C}^2} \qquad \text{5.69}$$

The distribution each of the terms is independently distributed as

$$H_1(z) = \frac{1}{\sqrt{2\pi}} z^{-\frac{1}{2}} e^{-\frac{z}{2}} \qquad \text{5.70}$$

The distribution of the sum of two terms is

$$H_2(z) = \frac{1}{2} e^{-\frac{z}{2}} \qquad \text{5.71}$$

The sum of three variables may be found from the convolution of H_1 and H_2

$$H_3(z) = \int_{-\infty}^{\infty} H_2(w) H_1(z - w) \, dw \qquad \text{5.72}$$

$$= \frac{1}{2\sqrt{2\pi}} \int_{-\infty}^{\infty} e^{-\frac{w}{2}} (z - w)^{-\frac{1}{2}} e^{-\frac{(z-w)}{2}} \, dw \qquad \text{5.73}$$

$$= \frac{1}{2\sqrt{2\pi}} z^{-1/2} e^{-\frac{z}{2}} \int_{-\infty}^{\infty} \left(1 - \frac{w}{z}\right)^{-\frac{1}{2}} \, dw \qquad \text{5.74}$$

$$= \frac{1}{2\sqrt{2\pi}} z^{1/2} e^{-\frac{z}{2}} \int_{-\infty}^{\infty} (1 - u)^{-\frac{1}{2}} \, du \qquad \text{5.75}$$

$$= \frac{1}{\sqrt{2\pi}} z^{1/2} e^{-\frac{z}{2}} \qquad \text{5.76}$$

This is a χ^2 distribution on three degrees of freedom. The probability that the statistic is larger than some value H by chance is

$$P_3(H_3 \geq H) = \frac{2}{\sqrt{\pi}} \gamma\left(\frac{3}{2}, \frac{H}{2}\right)$$ 5.77

5.4 Operations

Binary classifiers had operations for the dual and negation. The negation was realizable based on a classifier by negating the results. The dual was not realizable, meaning that we cannot construct the dual from the classifier.

The base matrix for the binary classifier is given by

$$B = \begin{pmatrix} \varphi & \chi \\ 1 - \varphi & 1 - \chi \end{pmatrix}$$ 5.78

The negated matrix is

$$\bar{B} = \begin{pmatrix} 1 - \varphi & 1 - \chi \\ \varphi & \chi \end{pmatrix}$$ 5.79

The negated matrix just swaps the rows of the base matrix. Similarly, the dual is

$$B^* = \begin{pmatrix} \chi & \varphi \\ 1 - \chi & 1 - \varphi \end{pmatrix}$$ 5.80

The dual swaps the columns of the base matrix. Finally, the negated dual is

$$\bar{B}^* = \begin{pmatrix} 1 - \chi & 1 - \varphi \\ \chi & \varphi \end{pmatrix}$$ 5.81

The negated dual swaps both the rows and the columns of the base matrix.

We see that the operations of negation and the dual swap rows and columns of the base matrix. We can extend this to the base matrix of the trinary classifier. Swapping rows A and B means that when the classifier says A, we interpret this as B, and when the classifier says B, we interpret this as A. Similarly, if we swap rows A and C, then every time the classifier outputs A, we assign the input to class C. Alternatively, when the classifier outputs C, we assign the input to class A. As row B remains in place, when the classifier outputs B, we assign the input to class B.

In fact, every permutation of the rows is a distinct operator. With three rows there are six total permutations. Swapping rows simply means that we reinterpret the results of the classifier in different ways. These operations are always realizable based on the results of the classifier.

Table 47 shows the six operations on the trinary classifier. Each of these operations is realizable because each instructs us to interpret the output of the classifier in different ways. We simply look at the output of the classifier, and map the output to a different class depending on which of the operations we are examining. Furthermore, because these operations are based on permuting the rows of the base matrix, these operations form a mathematical group.

Permutation	Interpretation
(1)	Identity permutation.
(AB)	Interpret A as B. Interpret B as A. C remains C.
(AC)	Interpret A as C. Interpret C as A. B Remains B.
(BC)	Interpret B as C. Interpret C as B. A remains A.
(ABC)	Interpret A as B. Interpret B as C. Interpret C as A.
(ACB)	Interpret A as C. Interpret C as B. Interpret B as B.

Table 47: Realizable operations on the trinary classifier.

We can carry out a similar procedure for the duals. For the duals, we swap the columns in the base matrix. Since the columns are independent of each other, we cannot realize a dual based on the output of the classifier.

There are six ways to permute three columns. Thus we have six duals we can identify. Moreover, we can combine the operations of duals and negations. Each dual may be negated in six different ways. This leads to 36 different matrices we can form from the base matrix. Again, only six of these are realizable from the results of the classifier. The other thirty are conceptually related, but not realizable from the classifier.

We can create a generating process to identify all of these matrices. Let $P_{(...)}$ represent a permutation on a row and let $P^{(...)}$ represent a permutation on a column. Then the list of matrices can be found from

$$\left(P_{(1)} + P_{(AB)} + P_{(AC)} + P_{(BC)} + P_{(ABC)} + P_{(ACB)}\right) \times$$
$$\left(P^{(1)} + P^{(AB)} + P^{(AC)} + P^{(BC)} + P^{(ABC)} + P^{(ACB)}\right) \tag{5.82}$$

Each term in the resulting expression is a distinct matrix and represents a different operation on the base matrix. Table 48 lists the results of applying every operator to the base matrix.

	$P_{(1)}$	$P_{(AB)}$	$P_{(AC)}$	$P_{(BC)}$	$P_{(ABC)}$	$P_{(ACB)}$
$P^{(1)}$	$\begin{pmatrix}\varphi_A & \chi_A & \psi_A\\ \varphi_B & \chi_B & \psi_B\\ \varphi_C & \chi_C & \psi_C\end{pmatrix}$	$\begin{pmatrix}\chi_A & \varphi_A & \psi_A\\ \chi_B & \varphi_B & \psi_B\\ \chi_C & \varphi_C & \psi_C\end{pmatrix}$	$\begin{pmatrix}\psi_A & \chi_A & \varphi_A\\ \psi_B & \chi_B & \varphi_B\\ \psi_C & \chi_C & \varphi_C\end{pmatrix}$	$\begin{pmatrix}\varphi_A & \psi_A & \chi_A\\ \varphi_B & \psi_B & \chi_B\\ \varphi_C & \psi_C & \chi_C\end{pmatrix}$	$\begin{pmatrix}\chi_A & \psi_A & \varphi_A\\ \chi_B & \psi_B & \varphi_B\\ \chi_C & \psi_C & \varphi_C\end{pmatrix}$	$\begin{pmatrix}\psi_A & \varphi_A & \chi_A\\ \psi_B & \varphi_B & \chi_B\\ \psi_C & \varphi_C & \chi_C\end{pmatrix}$
$P^{(AB)}$	$\begin{pmatrix}\varphi_B & \chi_B & \psi_B\\ \varphi_A & \chi_A & \psi_A\\ \varphi_C & \chi_C & \psi_C\end{pmatrix}$	$\begin{pmatrix}\chi_B & \varphi_B & \psi_B\\ \chi_A & \varphi_A & \psi_A\\ \chi_C & \varphi_C & \psi_C\end{pmatrix}$	$\begin{pmatrix}\psi_B & \chi_B & \varphi_B\\ \psi_A & \chi_A & \varphi_A\\ \psi_C & \chi_C & \varphi_C\end{pmatrix}$	$\begin{pmatrix}\varphi_B & \psi_B & \chi_B\\ \varphi_A & \psi_A & \chi_A\\ \varphi_C & \psi_C & \chi_C\end{pmatrix}$	$\begin{pmatrix}\chi_B & \psi_B & \varphi_B\\ \chi_A & \psi_A & \varphi_A\\ \chi_C & \psi_C & \varphi_C\end{pmatrix}$	$\begin{pmatrix}\psi_B & \varphi_B & \chi_B\\ \psi_A & \varphi_A & \chi_A\\ \psi_C & \varphi_C & \chi_C\end{pmatrix}$
$P^{(AC)}$	$\begin{pmatrix}\varphi_C & \chi_C & \psi_C\\ \varphi_B & \chi_B & \psi_B\\ \varphi_A & \chi_A & \psi_A\end{pmatrix}$	$\begin{pmatrix}\chi_C & \varphi_C & \psi_C\\ \chi_B & \varphi_B & \psi_B\\ \chi_A & \varphi_A & \psi_A\end{pmatrix}$	$\begin{pmatrix}\psi_C & \chi_C & \varphi_C\\ \psi_B & \chi_B & \varphi_B\\ \psi_A & \chi_A & \varphi_A\end{pmatrix}$	$\begin{pmatrix}\varphi_C & \psi_C & \chi_C\\ \varphi_B & \psi_B & \chi_B\\ \varphi_A & \psi_A & \chi_A\end{pmatrix}$	$\begin{pmatrix}\chi_C & \psi_C & \varphi_C\\ \chi_B & \psi_B & \varphi_B\\ \chi_A & \psi_A & \varphi_A\end{pmatrix}$	$\begin{pmatrix}\psi_C & \varphi_C & \chi_C\\ \psi_B & \varphi_B & \chi_B\\ \psi_A & \varphi_A & \chi_A\end{pmatrix}$
$P^{(BC)}$	$\begin{pmatrix}\varphi_A & \chi_A & \psi_A\\ \varphi_C & \chi_C & \psi_C\\ \varphi_B & \chi_B & \psi_B\end{pmatrix}$	$\begin{pmatrix}\chi_A & \varphi_A & \psi_A\\ \chi_C & \varphi_C & \psi_C\\ \chi_B & \varphi_B & \psi_B\end{pmatrix}$	$\begin{pmatrix}\psi_A & \chi_A & \varphi_A\\ \psi_C & \chi_C & \varphi_C\\ \psi_B & \chi_B & \varphi_B\end{pmatrix}$	$\begin{pmatrix}\varphi_A & \psi_A & \chi_A\\ \varphi_C & \psi_C & \chi_C\\ \varphi_B & \psi_B & \chi_B\end{pmatrix}$	$\begin{pmatrix}\chi_A & \psi_A & \varphi_A\\ \chi_C & \psi_C & \varphi_C\\ \chi_B & \psi_B & \varphi_B\end{pmatrix}$	$\begin{pmatrix}\psi_A & \varphi_A & \chi_A\\ \psi_C & \varphi_C & \chi_C\\ \psi_B & \varphi_B & \chi_B\end{pmatrix}$
$P^{(ABC)}$	$\begin{pmatrix}\varphi_B & \chi_B & \psi_B\\ \varphi_C & \chi_C & \psi_C\\ \varphi_A & \chi_A & \psi_A\end{pmatrix}$	$\begin{pmatrix}\chi_B & \varphi_B & \psi_B\\ \chi_C & \varphi_C & \psi_C\\ \chi_A & \varphi_A & \psi_A\end{pmatrix}$	$\begin{pmatrix}\psi_B & \chi_B & \varphi_B\\ \psi_C & \chi_C & \varphi_C\\ \psi_A & \chi_A & \varphi_A\end{pmatrix}$	$\begin{pmatrix}\varphi_B & \psi_B & \chi_B\\ \varphi_C & \psi_C & \chi_C\\ \varphi_A & \psi_A & \chi_A\end{pmatrix}$	$\begin{pmatrix}\chi_B & \psi_B & \varphi_B\\ \chi_C & \psi_C & \varphi_C\\ \chi_A & \psi_A & \varphi_A\end{pmatrix}$	$\begin{pmatrix}\psi_B & \varphi_B & \chi_B\\ \psi_C & \varphi_C & \chi_C\\ \psi_A & \varphi_A & \chi_A\end{pmatrix}$
$P^{(ACB)}$	$\begin{pmatrix}\varphi_C & \chi_C & \psi_C\\ \varphi_A & \chi_A & \psi_A\\ \varphi_B & \chi_B & \psi_B\end{pmatrix}$	$\begin{pmatrix}\chi_C & \varphi_C & \psi_C\\ \chi_A & \varphi_A & \psi_A\\ \chi_B & \varphi_B & \psi_B\end{pmatrix}$	$\begin{pmatrix}\psi_C & \chi_C & \varphi_C\\ \psi_A & \chi_A & \varphi_A\\ \psi_B & \chi_B & \varphi_B\end{pmatrix}$	$\begin{pmatrix}\varphi_C & \psi_C & \chi_C\\ \varphi_A & \psi_A & \chi_A\\ \varphi_B & \psi_B & \chi_B\end{pmatrix}$	$\begin{pmatrix}\chi_C & \psi_C & \varphi_C\\ \chi_A & \psi_A & \varphi_A\\ \chi_B & \psi_B & \varphi_B\end{pmatrix}$	$\begin{pmatrix}\psi_C & \varphi_C & \chi_C\\ \psi_A & \varphi_A & \chi_A\\ \psi_B & \varphi_B & \chi_B\end{pmatrix}$

Table 48: Trinary classifier operations and resulting matrices.

5.5 Types of Trinary Classifiers

5.5-a PERFECT CLASSIFIER

The confusion matrix for a perfect classifier only has non-zero elements on the diagonal. The diagonal elements correspond to the correct classification of the input by the classifier. In terms of the base metrics, the confusion matrix is

$$\begin{bmatrix} \rho_A & 0 & 0 \\ 0 & \rho_B & 0 \\ 0 & 0 & \rho_C \end{bmatrix} \tag{5.83}$$

We also have

$$\phi_A = \chi_B = \psi_C = 1 \tag{5.84}$$

5.5-b FALLACIOUS CLASSIFIER

The fallacious classifier has zero for every element on the diagonal. All off diagonal elements are non-zero. The base matrix is

$$\begin{bmatrix} 0 & \chi_A\rho_B & \psi_A\rho_C \\ \varphi_B\rho_A & 0 & \psi_B\rho_C \\ (1-\varphi_B)\rho_A & (1-\chi_A)\rho_B & 0 \end{bmatrix} \qquad 5.85$$

There are an infinite number of ways to create a fallacious classifier because each column has a parameter that can range from 0 to 1. This is in contrast to the perfect classifier where this is exactly one configuration of base parameters.

5.5-c αT Classifiers

αT classifiers have categories of the form: A, not A, and undetermined. Undetermined could be either A or not A. This classifier is essentially a binary classifier with an added category for 'undetermined'. αT classifiers are useful when there are inputs that the classifier is unable to analyze, but is able to differentiate these inputs.

5.5-d βT Classifiers

The βT classifier has the categories A, B, and C. This is the typical trinary classifier, and is the model used in this chapter. Each input must be assigned to exactly one class. In this respect, this is a parallel to the βB classifier, and suffers from the same problems.

5.5-e γT Classifiers

γT classifiers are the trinary analog of the γB classifier. Here, we have class A, B, and undetermined. These types of classifiers usually arise as an extension of a βB classifier when the classifier is able to recognize that the classifier is unable to process certain inputs.

5.5-f δT Classifiers

The δT classifier has categories A, B, and neither. This is slightly different than the previous trinary classifiers. There, undetermined is interpreted as 'unknown'. For the δT classifier, the final category is definitively 'neither A nor B', rather than 'unknown'.

5.5-g εT Classifiers

The εT classifier treats the final category slightly differently. Here, we have classes A, B, and *both* A and B. This type of classifier typically arises as an extension of a βB classifier when the classes are not mutually exclusive.

6 Multiclass Classifiers

6.1 Overview

Multiclass classifiers further generalize the binary and trinary classifiers. In this section we designate the classes of a multiclass classifier as Cl_i. This means we are referring to the ith class of the multiclass classifier.

The following sections are parallels of the sections from the treatment of the trinary classifier. In each section we generalize the results from the trinary classifier to k classes.

A test inputs set has a predetermined number of inputs assigned to each class. Let N be the total number of inputs, and let N_i be the number of inputs in the ith class. We define

$$\rho_i = \frac{N_i}{N} \tag{6.1}$$

as the proportion of inputs from each class. The sum of these over all classes is constrained by

$$\sum_{i=1}^{N} \rho_i = 1 \tag{6.2}$$

For shorthand we write

$$[\rho_i]_i = 1 \tag{6.3}$$

where the bracket means we sum over all values of from 1 to N, and the subscript on the bracket tells us which index we are summing over.

6.2 Metrics

The metrics for a multiclass classifier are generalizations of the trinary classifiers. Much of the presentation for the trinary classifier is reused here with the required additions to take this from three categories to an arbitrary number.

6.2-a CONFUSION MATRIX

A multiclass classifier on k elements has a $k \times k$ confusion matrix. We represent these elements as $\mathfrak{C}_{ij}$. Similarly, the base matrix is a $k \times k$ matrix, and we designate this as $\mathfrak{B}_{ij}$. The diagonal elements in these matrices are the successful classifications. The off-diagonal elements are cases where the classifier incorrectly classified the input.

The elements of the confusion matrix form rows according to the true value of the input, and rows according to the results of the classifier

$$\mathfrak{C} = \begin{pmatrix} \mathfrak{C}_{11} & \cdots & \mathfrak{C}_{1k} \\ \vdots & \ddots & \vdots \\ \mathfrak{C}_{k1} & \cdots & \mathfrak{C}_{kk} \end{pmatrix} \qquad 6.4$$

The sums of the columns are the number of test inputs for each class. Specifically,

$$\left[\mathfrak{C}_{ij}\right]_i = N_j = \rho_j N \qquad 6.5$$

The element of the base matrix are computed by dividing each element of the confusion matrix by the sum of the element in the same column:

$$\mathfrak{B}_{ij} = \frac{\mathfrak{C}_{ij}}{\rho_j N} \qquad 6.6$$

Alternatively, we can write the element of the confusion matrix in terms of the base matrix:

$$\mathfrak{C}_{ij} = \mathfrak{B}_{ij}\rho_j N \qquad 6.7$$

6.2-b CONSTRAINTS

The confusion matrix has k^2 elements. Each column is constrained by 6.5, which provides k constraints. This leaves $k(k-1)$ independent elements.

Each column contains $k-1$ variables, and these variables are independent from the other variables in the confusion matrix. However, the variables within a column are dependent on each other. These are distributed from the Dirichlet distribution

$$B(k, l, N; x, y) = \frac{\Gamma\left(\sum_{i=1}^N a_i\right)}{\prod_{i=1}^N \Gamma(a_i)} \left(1 - \sum_{i=1}^{N-1} x_i\right)^{a_N - 1} \prod_{i=1}^{N-1} x_i^{a_i - 1} \qquad 6.8$$

$$a_i > 0 \quad 0 \le x_i \le 1 \quad 0 \le y \le 0 \quad \sum_{i=1}^N a_i \le 1$$

The mean and variance are

$$\mu_i = \frac{a_i}{\sum_{j=1}^{N} a_j} \qquad \sigma_i^2 = \frac{a_i\left(\sum_{j=1}^{N} a_j - a_i\right)}{\left(\sum_{j=1}^{N} a_j\right)^2 \left(\sum_{j=1}^{N} a_j + 1\right)} \qquad 6.9$$

The distribution parameters may be written in terms of the base metrics as

$$a_i = \mathcal{B}_{ij}(\rho_j N - 1)$$

$$a_N = (\rho_N N - 1)\left(\sum_{i=1}^{N-1} \mathcal{B}_{iN} - 1\right) \qquad 6.10$$

There is one set of these variables for each column in the confusion matrix.

6.2-c N-AIRY METRICS

The multiclass classifier on k classes has $k(k-1)$ independent parameters. The parameters may be written in terms of the elements of the confusion matrix $\mathfrak{C}_{ij}$ or the base matrix $\mathcal{B}_{ij}$.

6.2-c(i) INPUT SET METRICS

The input set metrics identify the number of test inputs in each category Cl_i. Let N_j be the number of inputs in the jth class. The input metrics are

$$N_j = \left[\mathfrak{C}_{ij}\right]_i = \rho_j N \qquad 6.11$$

There are k such metrics, one for each column of the confusion (base) matrix.

6.2-c(ii) CLASSIFIED SET METRICS

The classified set metrics identify the number of inputs the classifier places into each class. These are the sums of the rows of the confusion matrix.

$$C^i = \left[\mathfrak{C}_{ij}\right]_j \qquad 6.12$$

or the base matrix

$$C^i = \left[\rho_j \mathcal{B}_{ij}\right]_j N \qquad 6.13$$

6.2-c(iii) CONFUSION MATRIX METRICS

The elements of the confusion matrix may be written in terms of the base matrix

$$\mathfrak{C}_{ij} = \mathcal{B}_{ij}\rho_j N \qquad 6.14$$

Alternatively, we may write the element of the base matrix in terms of the confusion matrix

$$\mathcal{B}_{ij} = \frac{\mathfrak{C}_{ij}}{\rho_j N} \qquad\qquad 6.15$$

6.2-c(iv) PROBABILITY METRICS

The probability metrics for the multiclass classifier generalize those for the trinary classifier. In general we have probability metrics with the forms $P\left(\bigvee_{i=1}^{n} T_{k_i} \mid \bigvee_{i=1}^{m} C_{l_i}\right)$ and $P\left(\bigvee_{i=1}^{m} C_{l_i} \mid \bigvee_{i=1}^{n} T_{k_i}\right)$.

There are 2^k total events of the form $\bigvee_{i=1}^{n} T_{k_i}$, and the same number with the form $\bigvee_{i=1}^{m} C_{l_i}$. One of these forms has all possible events present, and another has no events present. The form with all possible events is uninteresting because when we 'or' all events, the result is just 1. The form with no events is 0.

If we remove the uninteresting cases, we are left with $2^k - 2$ cases for each side of the $|$. We can create a metric for each combination from either side of the $|$. This yields $2^{2k} - 2^{k+2} + 4$ different probability metrics for $P\left(\bigvee_{i=1}^{n} T_{k_i} \mid \bigvee_{i=1}^{m} C_{l_i}\right)$ and the same number for $P\left(\bigvee_{i=1}^{m} C_{l_i} \mid \bigvee_{i=1}^{n} T_{k_i}\right)$. The total number of probability metrics is then $2^{2k+1} - 2^{k+3} + 8$.

As a simple test, if we set $k = 2$, our formula predicts 8 probability metrics, and with $k = 3$, we find 72 metrics. This agrees with our analysis from 5.2-c(iv).

Each of these metrics takes the form of a ratio. The different T's determine a set of columns in the confusion matrix, while the C's determine a set of rows. The numerator of the ratio is the sum of the cells of the confusion matrix where the rows and columns intersect. The denominator is determined by the events on the right side of the $|$. If the events on the right are T's, the denominator is the sum of the T columns. If the events on the right are C's, the denominator is the sum of the C columns.

Each of the probability metrics provides insight to the performance of the classifier. The metrics $P(T_k|C_l)$ and $P(C_k|T_l)$ are particularly useful. These metrics are given by

$$P(T_k|C_l) = \frac{\mathfrak{C}_{lk}}{[\mathfrak{C}_{li}]_i} \qquad\qquad 6.16$$

$$P(C_k|T_l) = \frac{\mathfrak{C}_{kl}}{[\mathfrak{C}_{ik}]_i} \qquad\qquad 6.17$$

In terms of the base metrics there are

$$P(T_k|C_l) = \frac{\mathcal{B}_{lk}\rho_k}{[\mathcal{B}_{li}\rho_i]_i} \qquad 6.18$$

$$P(C_k|T_l) = \mathcal{B}_{kl} \qquad 6.19$$

More generally, $P\left(V_{i=1}^n T_{k_i} \mid V_{i=1}^m C_{l_i}\right)$ and $P\left(V_{i=1}^m C_{l_i} \mid V_{i=1}^n T_{k_i}\right)$ are

$$P\left(\bigvee_{i=1}^n T_{k_i} \mid \bigvee_{i=1}^m C_{l_i}\right) = \frac{\Sigma_{k_i=l_i}\mathfrak{C}_{l_i k_i}}{\Sigma_{i=1}^m [\mathfrak{C}_{l_i j}]_j} \qquad 6.20$$

$$P\left(\bigvee_{i=1}^m C_{l_i} \mid \bigvee_{i=1}^n T_{k_i}\right) = \frac{\Sigma_{k_i=l_i}\mathfrak{C}_{l_i k_i}}{\Sigma_{i=1}^n [\mathfrak{C}_{j k_i}]_j} \qquad 6.21$$

In terms of the base metrics

$$P\left(\bigvee_{i=1}^n T_{k_i} \mid \bigvee_{i=1}^m C_{l_i}\right) = \frac{\Sigma_{k_i=l_i}\mathcal{B}_{l_i k_i}\rho_{k_i}}{\Sigma_{i=1}^m [\mathcal{B}_{l_i j}\rho_j]_j} \qquad 6.22$$

$$P\left(\bigvee_{i=1}^m C_{l_i} \mid \bigvee_{i=1}^n T_{k_i}\right) = \frac{\Sigma_{k_i=l_i}\mathcal{B}_{l_i k_i}\rho_{k_i}}{\Sigma_{i=1}^n \rho_{k_i}} \qquad 6.23$$

6.2-c(v) ACCURACY

Accuracy is the ratio of the trace of the confusion matrix divided by the total number of inputs:

$$ACC = \frac{[\mathfrak{C}_{ii}]_i}{N} \qquad 6.24$$

In terms of the base metrics,

$$ACC = [\mathcal{B}_{ii}\rho_i]_i \qquad 6.25$$

6.2-c(vi) DETERMINANT

The determinant metric is

$$D = \frac{|\mathfrak{C}|}{N^k} \qquad 6.26$$

6.3 Error Analysis

For comparing two multiclass classifiers, we examine the positive performance matrix. For multiclass classifiers with k classes, the positive performance matrix has k columns and two rows. The elements of the first row are the diagonal elements of the base matrix. The elements in the second row are one minus the elements from the first row:

$$PPM = \begin{pmatrix} \mathcal{B}_{11} & \cdots & \mathcal{B}_{kk} \\ 1 - \mathcal{B}_{11} & \cdots & 1 - \mathcal{B}_{kk} \end{pmatrix} \qquad 6.27$$

Each of these variables is Beta distributed and independent of the other variables. The variance for each variable is

$$\sigma^2_{\mathcal{B}_{ii}} = \frac{\mathcal{B}_{ii}(1 - \mathcal{B}_{ii})}{\rho_i N} \qquad 6.28$$

Again, we are faced with the question of which classifier is better. For multiclass classifiers, this question is almost never answerable without a tradeoff function. As the number of categories increases, the likelihood that we will find two different classifiers where one has every performance metric greater than the other becomes increasingly unlikely. As the number of classes grows large, we will almost certainly find at least one category from one classifier that outperforms the second classifier, and another category where the second outperforms the first.

In the event where we do have all base metrics of one classifier outperforming the other, we can use Monte Carlo, simulation, and decomposition into binary classifiers to compare the classifiers and determine if one is statistically better than the other.

In the situation where the classifier metrics are incomparable and there is no suitable tradeoff function, we appeal to other techniques to compare classifier performance.

There is no general method for objectively determining relative performance in these cases. Often an understanding of the underlying application of the classifiers to a specific problem is used to construct an appropriate metric. Again, this is another version of a tradeoff function.

Although there is no general method of selecting between these classifiers, the next sections provide some examples. These examples are general and do not take into account any features of the specific problem treated by the classifiers. However, in many cases, these metrics are useful in understanding the relationship between the classifiers.

6.3-a ACCURACY

The accuracy metric is a simple metric indicating the overall accuracy of a classifier. The accuracy is weighted according to the input set ratios. If the specific problem expects different ratios between the classes, we can modify the accuracy. We use the same formula

$$ACC = [\mathcal{B}_{ii}\rho_i]_i \qquad\qquad 6.29$$

except we set ρ_i as the expected proportion of the ith class rather than the proportion seen in the test set. This will provide an estimate of the expected accuracy for the classifier when applied to the specific problem. By comparing these values we can identify which classifier is likely to have the higher overall accuracy.

6.3-b POSITIVE PERFORMANCE MATRIX

Alternatively, we can use a weighted average of the first row from the positive performance matrix. The first row is the same as the diagonal of the base matrix, which are the same base metrics that comprise the accuracy metric.

Here, we weight each metric by the overall value we assign to correctly classifying the class. If these weights are simply the expected proportion of an input belonging to a particular class, then this is equivalent to the accuracy metric from the last section.

We do not need to use the expected proportions as the weights. Instead, we may assign a value to correctly classifying for each category. For example, we might be more interested in correctly identifying defects, even though we expect a far greater of inputs are non-defective.

The overall formula for this metric is the same as for accuracy:

$$P = [\mathcal{B}_{ii}\rho_i]_i \qquad\qquad 6.30$$

The difference is in the interpretation of the weights ρ_i.

6.3-c CLASSIFICATION SIGNIFICANCE

If we treat the multiclass classifiers as a set of binary classifiers, we can compare similar binary classifiers between the two multiclass classifiers. For example, we can compare the performance of the binary classifier associated with class A from the first multiclass classifier against the binary classifier associated with class A from the second multiclass classifier.

If there are k classes, we can make k such comparisons. For each comparison, we can determine if the first multiclass classifier significantly outperforms the second, if the second outperforms the first, or if the two are statistically within error.

From this we construct a table counting the number of binary classifiers that significantly outperform. We have three entries: count of first outperforming the second, count of statistical ties, count of second outperforming the first.

Based on this we select the classifier that has the most binary subcomponents that perform well.

6.3-d DECOMPOSITION TO BINARY CLASSIFIERS

If the multiclass classifier is treated as a set of binary classifiers, the performance metrics for each is independent, and we may use Hotelling's T-Square to compute a useful statistic.

From section 2.7-c we found the distribution of the square of the difference in metrics divided by the variance of the difference. Applying this to multiclass classifiers, we examine the statistic

$$H = \sum_{i=1}^{k} \frac{\left(\mathcal{B}_{ii} - \bar{\bar{\mathcal{B}}}_{ii}\right)^2}{\sigma_{\mathcal{B}_{ii}}^2 + \sigma_{\bar{\bar{\mathcal{B}}}_{ii}}^2} \qquad 6.31$$

From section 2.7-c, the distribution of the sum with only one term is

$$H_1(z) = \frac{1}{\sqrt{2\pi}} z^{-\frac{1}{2}} e^{-\frac{z}{2}} \qquad 6.32$$

The distribution of the sum with two terms is

$$H_2(z) = \frac{1}{2} e^{-\frac{z}{2}} \qquad 6.33$$

And from section 5.3-c

$$H_3(z) = \frac{1}{\sqrt{2\pi}} z^{1/2} e^{-\frac{z}{2}} \qquad 6.34$$

Each of these is a χ^2 distribution. The first is on one degree of freedom, the second on two degrees, and the third on three degrees. This leads us to suspect that the sum of n such variables might be a χ^2 distribution with n-degrees of freedom. We prove this is true by induction.

The χ^2 distribution on n-degrees of freedom is

$$\chi^2(n; z) = \frac{1}{2^{n/2} \Gamma(n/2)} z^{\frac{n}{2}-1} e^{-\frac{z}{2}} \qquad 6.35$$

Suppose

$$H_n(z) = \chi^2(n; z) \qquad 6.36$$

Then the sum on $n + 1$ such variables is given by the convolution of this with H_1

$$H_{n+1}(z) = \int_{-\infty}^{\infty} H_n(w) H_1(z-w) \, dw \qquad 6.37$$

$$= \frac{1}{2^{n/2}\sqrt{2\pi}\,\Gamma(n/2)} \int_{-\infty}^{\infty} w^{\frac{n}{2}-1} e^{-\frac{w}{2}} (z-w)^{-\frac{1}{2}} e^{-\frac{(z-w)}{2}} \, dw \qquad 6.38$$

$$= \frac{1}{2^{n/2}\sqrt{2\pi}\,\Gamma(n/2)} e^{-\frac{z}{2}} \int_{-\infty}^{\infty} w^{\frac{n}{2}-1} (z-w)^{-\frac{1}{2}} \, dw \qquad 6.39$$

$$= \frac{1}{2^{n/2}\sqrt{2\pi}\,\Gamma(n/2)} z^{-\frac{1}{2}} e^{-\frac{z}{2}} \int_{-\infty}^{\infty} w^{\frac{n}{2}-1} \left(1 - \frac{w}{z}\right)^{-\frac{1}{2}} \, dw \qquad 6.40$$

$$= \frac{1}{2^{n/2}\sqrt{2\pi}\,\Gamma(n/2)} z^{\frac{n-3}{2}} e^{-\frac{z}{2}} \int_{-\infty}^{\infty} \left(\frac{w}{z}\right)^{\frac{n}{2}-1} \left(1 - \frac{w}{z}\right)^{-\frac{1}{2}} \, dw \qquad 6.41$$

$$= \frac{1}{2^{n/2}\sqrt{2\pi}\,\Gamma(n/2)} z^{\frac{n-1}{2}} e^{-\frac{z}{2}} \int_{-\infty}^{\infty} u^{\frac{n}{2}-1} (1-u)^{-\frac{1}{2}} \, du \qquad 6.42$$

$$= \frac{1}{2^{(n+1)/2}\,\Gamma((n+1)/2)} z^{\frac{n+1}{2}-1} e^{-\frac{z}{2}} \qquad 6.43$$

$$= \chi^2(n+1; z) \qquad 6.44$$

Thus, if H_n is a χ^2 distribution on n-degrees of freedom, then H_{n+1} is a χ^2 distribution on $n+1$-degrees of freedom. Since we know that H_1 is a χ^2 distribution with 1 degree of freedom, the proof by induction is complete.

From this, the probability that the statistic is larger than H by chance is

$$P_n(H_n \geq H) = \frac{1}{\Gamma\left(\frac{n}{2}\right)} \gamma\left(\frac{n}{2}, \frac{x}{2}\right) \qquad 6.45$$

$$= P\left(\frac{n}{2}, \frac{x}{2}\right) \qquad 6.46$$

where $P(n, x)$ is the regularized gamma function. Tables of the regularized gamma function are provided in Appendix E.

The Hotelling statistic is useful when comparing multiclass classifiers. The statistic is easy to compute and may be compared with tables of the lower incomplete gamma function to determine the level of significance.

The Hotelling statistic determines if two classifiers are statistically different. It does not tell us which is better, only that the two are different. When assessing classifier performance, we use this test in conjunction with the others in this section to identify exactly how two classifiers compare.

6.4 Operations

Operations for the multiclass classifier are analogous to the negation and duals for the binary and trinary classifier. Interchanging the rows or columns of the confusion matrix leads to different operators on the classifier.

There are $k!$ Different combinations for each of the rows and columns. Thus, there are $k!^2$ total operations on a classifier with k categories.

Operations that only exchange rows are realizable. These cases take the results of the classifier and map them to a different value. For instance, when the classifier says A, we interpret this a B. There are $k!$ Total operations that only exchange rows. These are analogs of the negation operator from the binary classifier.

6.5 Types of Multiclass Classifiers

The ROC space for the multiclass classifier has $k(k-1)$ dimensions. The number of dimensions in the ROC space is equal to the number of independent parameters.

As the number of dimensions increases, the number of variants of types of multiclass classifiers increase. Typically, multiclass classifiers has a set of distinct categories, and possibly a category for 'None of the Above'.

Combination classifiers allow each input to be placed in more than one category. Combination classifiers can be made into pigeonhole classifiers by adding a new class for each combination of the original classifiers.

However, this approach quickly increases the number of categories. For example, if we begin with a set of k classes, there are 2^k combinations of these categories.

Another perspective is that a multiclass classifier is a series of binary classifiers, with one binary classifier for every class in the multiclass classifier. We can examine each of these binary classifiers separately using the binary classifier metrics.

The variety of multiclass classifiers may be analyzed using the performance metrics identified in this and the previous chapters. The various metrics each provide a different insight to the capabilities of the classifier.

7 Applications of Classifiers

In this chapter we examine application of classifiers. As an example we use a classifier that reviews documents and determines the language of the document. We use the tools developed previously to analyze the performance of the classifier in different situations.

The classifier is based on a set of algorithms. There are two main algorithms. The first examines the words in the document, and the second examines the letters. Each of these is a classifier in itself, and each makes a prediction of language. The final classifier combines these results using a combining algorithm to make a final prediction.

The combining algorithm has several variants as well. Each variant of the combining algorithm produces slightly different classifier. The performance metrics may be used to compare the different variants of the final classifier.

7.1 Binary Classifier

As a binary classifier, we present the classifier with a document and the classifier determines if the document is in English or not. The categories for the classifier are 'English' or 'not English'.

For the test input set, we choose 2000 documents, where 1000 are English, and 1000 are not English. We present these to the classifier and find the confusion matrix in Table 49.

	Confusion Matrix		Base Matrix	
	Actual English	Actual Not English	Actual English	Actual Not English
Classified English	918	27	.918	.027
Clarified Not English	82	973	.082	.973

Table 49: Confusion and Base matrices for a binary language classifier.

From Table 19, the distribution of the base metrics is

$$\mathfrak{D}_I(\alpha, \beta, \lambda; z) = \frac{z^{\alpha-1}(1-z)^{\beta-1}}{B(\alpha, \beta)} \tag{7.1}$$

and from Table 20 the parameters are

$$\begin{aligned} \alpha &= \varphi(\rho N - 1) \\ \beta &= (1 - \varphi)(\rho N - 1) \\ \lambda &= \rho N \end{aligned} \tag{7.2}$$

for the distribution of φ, and

$$\begin{aligned} \alpha &= \chi\big((1-\rho)N - 1\big) \\ \beta &= (1 - \chi)\big((1-\rho)N - 1\big) \\ \lambda &= (1-\rho)N \end{aligned} \tag{7.3}$$

for χ. From the input numebrs and Table 49 we have

$$\begin{aligned} N &= 2000 \\ \rho &= 0.5 \\ \varphi &= 0.918 \\ \chi &= 0.027 \end{aligned} \tag{7.4}$$

Substituting these into the distribution parameters,

$$\begin{aligned} \alpha &= 917 \\ \beta &= 82 \\ \lambda &= 1000 \end{aligned} \tag{7.5}$$

for the φ distribution and

$$\begin{aligned} \alpha &= 27 \\ \beta &= 927 \\ \lambda &= 1000 \end{aligned} \tag{7.6}$$

for the χ distribution. In these cases, $N > 50$ and $\varphi N > 5$ and $\chi N > 5$, so the conditions from Table 9 meaning we can approximate these distributions as Normal.

If we approximate these distributions with a Normal distribution, the variance is estimated from equations 2.124 and 2.125:

$$\sigma_\varphi = \sqrt{\frac{\varphi(1-\varphi)}{\rho N}} = 0.0087 \tag{7.7}$$

$$\sigma_\chi = \sqrt{\frac{\chi(1-\chi)}{(1-\rho)N}} = 0.0051 \tag{7.8}$$

The overall results of this run of the test input set is

$$\varphi = 0.918 \quad \sigma_\varphi = 0.0087$$
$$\chi = 0.027 \quad \sigma_\chi = 0.0051$$

7.9

Under a different configuration of the classifier, we find the confusion matrix

	Confusion Matrix		Base Matrix	
	Actual English	Actual Not English	Actual English	Actual Not English
Classified English	949	17	.949	.017
Clarified Not English	51	983	.051	.983

Table 50: Confusion and Base matrices for a binary language classifier.

This leads to the metrics

$$\varphi = 0.949 \quad \sigma_\varphi = 0.0070$$
$$\chi = 0.017 \quad \sigma_\chi = 0.0041$$

7.10

We can plot these points in ROC space. This plot is shown in Figure 60. The error bounds on the points overlap slightly. The z-score between these points is determined using 2.132.

$$z = \frac{(\varphi - \bar{\varphi})^2 + (\chi - \bar{\chi})^2}{\sqrt{(\varphi - \bar{\varphi})^2 (\sigma_\varphi^2 + \sigma_{\bar{\varphi}}^2) + (\chi - \bar{\chi})^2 (\sigma_\chi^2 + \sigma_{\bar{\chi}}^2)}} = 3.02$$

7.11

These points are almost at the 99.8% confidence level for difference. This statistic finds the points are significantly different, and the second classifier outperforms the first.

The statistic itself does not tell us that the second classifier outperforms the first. The statistic only tells us that the two performance metrics are unlikely to result from the same classifier. In this sense, the classifiers must be different. We know that the second classifier outperforms the first is by observing that the second classifier has better values for both performance metrics.

Next, examine the sum of the absolute value of the z-scores. This statistic is tabulated in Appendix E.

$$K = \left| \frac{\varphi - \bar{\varphi}}{\sqrt{\sigma_\varphi^2 + \sigma_{\bar{\varphi}}^2}} \right| + \left| \frac{\chi - \bar{\chi}}{\sqrt{\sigma_\chi^2 + \sigma_{\bar{\chi}}^2}} \right| = 4.30$$

7.12

Using the table for Absolute Z-Score Sum in Appendix E, with $N = 2$, the value of the statistic corresponds to over 99.5% confidence in statistical difference. This is similar to the results obtained from the z-score distance statistic.

Furthermore, we can examine the Hotelling statistic. Here we examine the sum of the squares of the z-scores

$$H = \frac{(\varphi - \bar{\varphi})^2}{\sigma_\varphi^2 + \sigma_{\bar{\varphi}}^2} + \frac{(\chi - \bar{\chi})^2}{\sigma_\chi^2 + \sigma_{\bar{\chi}}^2} = 10.04 \qquad 7.13$$

The probability the Hotelling statistic is greater than or equal to this value is given by

$$P_2(H_2 \geq H) = \gamma\left(1, \frac{H}{2}\right) \qquad 7.14$$

$$= \gamma(1, 5.02) \qquad 7.15$$

$$= .993 \qquad 7.16$$

The value of the lower incomplete gamma function can be found by examining the tables for the regularized gamma function in Appendix E. The likelihood of two classifiers producing these results by chance is less than 0.7%.

All three statistics indicate that these classifiers are different at about the 99% confidence level. Since the second classifier outperforms the first, we conclude that the second classifier statistically outperforms the first.

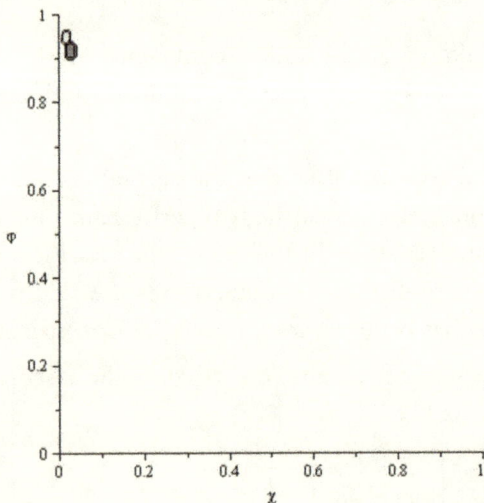

Figure 60: Two binary classifiers for language detection in RC space.

7.2 Trinary Classifier

For a trinary classifier, we expand the language identification to three categories: English, French, and Neither. This is a NOTA trinary classifier. The confusion and base matrices for the configuration using only words is provided in Table 51.

Word	Confusion Matrix			Base Matrix		
	Actual English	Actual French	Actual Neither	Actual English	Actual French	Actual Neither
Classified English	919	27	1	.919	.027	.001
Clarified French	19	914	2	.019	.914	.002
Classified Neither	62	59	997	.062	.059	.997

Table 51: Confusion and Base matrices for a trinary language classifier. The input set is 3000 documents, 1000 English, 1000 French, and 1000 Neither.

The base metrics for French and English satisfy $N > 50$, $\varphi_E \rho_E N > 5$, $\varphi_E \rho_E N > 5$, $\varphi_F \rho_E N > 5$, $\chi_E \rho_F N > 5$, and $\chi_F \rho_F N > 5$. Here, φ is the variable associated with the 'Actual English' columns, χ is associated with the 'Actual French' column, the subscript 'E' indicates 'Classified English' and the subscript 'F' indicates 'Classified French'.

Letter	Confusion Matrix			Base Matrix		
	Actual English	Actual French	Actual Neither	Actual English	Actual French	Actual Neither
Classified English	929	0	0	.929	0.0	0.0
Clarified French	5	753	70	.005	.753	.070
Classified Neither	66	247	930	.066	.247	.930

Table 52: Confusion and Base matrices for a trinary language classifier. The input set is 3000 documents, 1000 English, 1000 French, and 1000 Neither.

However, the 'Neither' column does not satisfy $\psi_E \rho_N N > 5$ or $\psi_F \rho_N N > 5$. These metrics are not well approximated with the normal distribution.

We want to compare the word version of the language classifier with the letter version. The metrics for the letter version are shown in Table 52.

The positive predictive matrices for these classifiers are

$$\textbf{\textit{Word}} \qquad \begin{pmatrix} .919 & .914 & .997 \\ .081 & .086 & .003 \end{pmatrix} \qquad\qquad 7.17$$

$$\textbf{\textit{Letter}} \qquad \begin{pmatrix} .929 & .753 & .930 \\ .071 & .247 & .070 \end{pmatrix} \qquad\qquad 7.18$$

Based on these numbers alone, the letter classifier outperforms the word classifier for English, but the word classifier outperforms on both French and Neither. We test each of these using simulation.

We setup the simulation to repeat 10,000 times. The results are

$$\begin{array}{ll} \varphi_{Word} > \varphi_{Letter} & .21 \\ \chi_{Word} > \chi_{Letter} & > .99 \\ \psi_{Word} > \psi_{Letter} & > .99 \end{array} \qquad\qquad 7.19$$

The φ are statistically tied from the perspective of the 95% confidence level because the probability is neither $> .95$ (Word > Letter) or $< .05$ (Letter > Word). However, both χ and ψ have the Word classifier outperforming the Letter classifier at the 95% confidence level.

Based on this, we conclude that the word classifier overall outperforms the letter classifier. These are statistically tied on one metrics, while the Word classifier significantly outperforms on the other metrics.

7.3 Trinary Classifier as Three Binary Classifiers

Let's examine the previous trinary classifier from the perspective of three binary classifiers. The base matrices for the word classifier are

$$\begin{matrix} \textbf{\textit{Word}} \\ \textbf{\textit{English}} \end{matrix} \qquad \begin{pmatrix} .919 & .014 \\ .081 & .986 \end{pmatrix} \qquad\qquad 7.20$$

$$\begin{matrix} \textbf{\textit{Word}} \\ \textbf{\textit{French}} \end{matrix} \qquad \begin{pmatrix} .914 & .0105 \\ .086 & .9895 \end{pmatrix} \qquad\qquad 7.21$$

$$\begin{matrix} \textbf{\textit{Word}} \\ \textbf{\textit{Neither}} \end{matrix} \qquad \begin{pmatrix} .997 & .121 \\ .003 & .879 \end{pmatrix} \qquad\qquad 7.22$$

Similarly, the base matrices for the letter classifier is

$$
\begin{matrix} \textit{Letter} \\ \textit{English} \end{matrix} \quad \begin{pmatrix} .929 & 0.0 \\ .071 & 1.0 \end{pmatrix} \tag{7.23}
$$

$$
\begin{matrix} \textit{Letter} \\ \textit{French} \end{matrix} \quad \begin{pmatrix} .753 & .0375 \\ .247 & .9625 \end{pmatrix} \tag{7.24}
$$

$$
\begin{matrix} \textit{Letter} \\ \textit{Neither} \end{matrix} \quad \begin{pmatrix} .930 & .1565 \\ .070 & .8435 \end{pmatrix} \tag{7.25}
$$

The variances for the binary classifiers is

$$
\begin{matrix} \textbf{\textit{Word}} \\ \textbf{\textit{English}} \end{matrix} \quad
\begin{aligned}
\sigma_\varphi^2 &= \frac{\varphi(1-\varphi)}{N} = 7.44 \times 10^{-5} \\
\sigma_\chi^2 &= \frac{\chi(1-\chi)}{N} = 1.38 \times 10^{-5}
\end{aligned} \tag{7.26}
$$

$$
\begin{matrix} \textbf{\textit{Word}} \\ \textbf{\textit{French}} \end{matrix} \quad
\begin{aligned}
\sigma_\varphi^2 &= \frac{\varphi(1-\varphi)}{N} = 7.86 \times 10^{-5} \\
\sigma_\chi^2 &= \frac{\chi(1-\chi)}{N} = 1.04 \times 10^{-5}
\end{aligned} \tag{7.27}
$$

$$
\begin{matrix} \textbf{\textit{Word}} \\ \textbf{\textit{Neither}} \end{matrix} \quad
\begin{aligned}
\sigma_\varphi^2 &= \frac{\varphi(1-\varphi)}{N} = 2.99 \times 10^{-6} \\
\sigma_\chi^2 &= \frac{\chi(1-\chi)}{N} = 1.06 \times 10^{-4}
\end{aligned} \tag{7.28}
$$

$$
\begin{matrix} \textbf{\textit{Letter}} \\ \textbf{\textit{English}} \end{matrix} \quad
\begin{aligned}
\sigma_\varphi^2 &= \frac{\varphi(1-\varphi)}{N} < 6.60 \times 10^{-5} \\
\sigma_\chi^2 &= \frac{\chi(1-\chi)}{N} < 1.00 \times 10^{-7}
\end{aligned} \tag{7.29}
$$

$$
\begin{matrix} \textbf{\textit{Letter}} \\ \textbf{\textit{French}} \end{matrix} \quad
\begin{aligned}
\sigma_\varphi^2 &= \frac{\varphi(1-\varphi)}{N} = 1.86 \times 10^{-4} \\
\sigma_\chi^2 &= \frac{\chi(1-\chi)}{N} = 3.61 \times 10^{-5}
\end{aligned} \tag{7.30}
$$

$$
\begin{matrix} \textbf{\textit{Letter}} \\ \textbf{\textit{Neither}} \end{matrix} \quad
\begin{aligned}
\sigma_\varphi^2 &= \frac{\varphi(1-\varphi)}{N} = 6.51 \times 10^{-5} \\
\sigma_\chi^2 &= \frac{\chi(1-\chi)}{N} = 1.32 \times 10^{-4}
\end{aligned} \tag{7.31}
$$

We can use the above values to compute the z-score for the distance between the classifiers. This is

$$
z = \frac{(\varphi - \bar{\varphi})^2 + (\chi - \bar{\chi})^2}{\sqrt{(\varphi - \bar{\varphi})^2(\sigma_\varphi^2 + \sigma_{\bar{\varphi}}^2) + (\chi - \bar{\chi})^2(\sigma_\chi^2 + \sigma_{\bar{\chi}}^2)}} \tag{7.32}
$$

$$z_\varphi = 2.09$$
$$z_\chi = 19.38$$
$$z_\psi = 11.48$$

At the 95% confidence level, we are comparing the z-score to the value 1.96. All three of these results are significant. This means that the English binary classifier of the Letter classifier is significantly better than the English classifier for the Word classifier. Moreover, the Word classifier has far better French and Neither components.

This result is slightly different than in the previous section. There we found the φ value for the Letter classifier was better, but it was not significant. Here, the English binary classifier for the Letter classifier is just over the 95% confidence level of significance.

The difference here is that with the binary classifier comparison, we are comparing the overall performance of the binary classifier. This means we are looking at both the φ and χ values. Because the χ for the English-Letter classifier is also better, the combination with the φ value pushes the result from just under significance to just over significance.

Based on these results, we would determine that the Letter and Word classifiers are overall not comparable. The Letter classifier has a significantly better English component, but the Word classifier has significantly better French and Neither classifiers.

This analysis is at the 95% confidence level. If we move to the 99% confidence level, then the z-score we compare is 2.58. At this level, the English classifiers are no longer significantly different, but the French and Neither classifier remain significantly different. At the 99% confidence level, the English classifiers are statistically tied, while the French and Neither classifier outperform in the Word classifier. In this case, we would determine that the Word classifier outperforms the Letter classifier.

Next we compute the Hotelling statistic. This statistic is given by

$$H_3 = \frac{(\varphi_A - \bar{\varphi}_A)^2}{\sigma_{\varphi_A}^2 + \sigma_{\bar{\varphi}_A}^2} + \frac{(\chi_B - \bar{\chi}_B)^2}{\sigma_{\chi_B}^2 + \sigma_{\bar{\chi}_B}^2} + \frac{(\psi_C - \bar{\psi}_C)^2}{\sigma_{\psi_C}^2 + \sigma_{\bar{\psi}_C}^2}$$

$$H_3 = 260.6$$

The corresponding confidence level is near unity. These classifiers are almost surely distinct. This agrees with the previous statistic where we found z-scores in excess of 10 for two components. This statistic is good for determining overall differences, but does not recognize the significance of each component.

7.4 Multiclass with Few Categories

We extend the classifier to include sixteen languages. Each language is a category for the multiclass classifier. We test the classifier using 1000 documents from each of the 16 languages.

The resulting confusion matrix has 256 entries. The multiclass classifier is becoming difficult to analyze via the full confusion matrix. At this point we begin examining indirect techniques to analyze the performance of two different classifiers.

The classifier is composed of both a word and letter classifier. The results are combined together using a combining algorithm. In this section we compare two different combining algorithms to see if one outperforms the other.

The diagonal elements of the base matrices are provided in Table 53.

Language	Combination 1	Combination 2
Amharic	.994	.992
Arabic	.990	.998
German	.995	.948
Spanish	.952	.905
French	.916	.924
Italian	.928	.939
Japanese	.999	.996
Korean	.924	.992
Dutch	.867	.855
Norwegian	.910	.897
Polish	.950	.943
Russian	.831	.893
English	.918	.951
Swedish	.897	.963
Ukrainian	.960	.956
Chinese	.023	.733

Table 53: Diagonal elements of the base matrices for two multiclass classifiers.

We begin by comparing the overall accuracy of each of the classifiers. We compute these from the diagonal elements of the base matrix. These accuracies are shown in Table 54.

	Combination 1	Combination 2
ACC	.878	.928

Table 54: Overall accuracy for two multiclass classifiers.

The overall accuracy of Combination 2 is higher than Combination 1 by .05. This indicates that 2 may be preferred to 1. From the diagonal elements of the base matrix, we see that much of this difference arises because Combination 2 has a 73.3% accuracy in Chinese, while Combination 1 only has 2.3% accuracy. This alone accounts for .044 of the difference between these. This amounts to nearly 90% of the difference between these classifiers.

Next, we examine the classifiers as a set of binary classifiers. There are sixteen such binary classifiers for each of the multiclass classifiers. We compare the metrics for each, and count the number that are statistically significant at both 95% and 99% confidence intervals. These results are shown in Table 55.

	1 > 2	Tie	2 > 1
95% Confidence	3	9	4
99% Confidence	2	11	3

Table 55: Count of significant differences between component binary classifiers between two multiclass classifiers.

At both the 95% and 99% levels, Combination 1 has one less significant binary classifier than Combination 2. The accuracy for 2 is better than 1, but this is mainly due to the performance in Chinese. Overall, 2 appears to be an improvement over 1, but the majority of this is due to Chinese.

We examine the Hotelling statistic to determine if these classifiers are truly distinct. This statistic is the sum of the squares of the z-scores for each of the component binary classifiers:

$$H = \sum_{i=1}^{k} \frac{\left(\mathcal{B}_{ii} - \bar{\bar{\mathcal{B}}}_{ii}\right)^2}{\sigma_{\mathcal{B}_{ii}}^2 + \sigma_{\bar{\bar{\mathcal{B}}}_{ii}}^2} \approx 2500 \qquad 7.36$$

There are sixteen degrees of freedom on this distribution. The result is a confidence level near unity. If we remove Chinese, then $H \approx 200$ with fifteen degrees of freedom. Still, the confidence that these are different classifiers is near unity.

Appendix A: Special Functions

A.1 Gamma Function

The Gamma function is a continuous extension of the factorial:

$$\Gamma(n) = \int_0^\infty z^{n-1}e^{-z}dz \qquad \text{A-1}$$

The Gamma function has the recurrence relation

$$\Gamma(n+1) = n\Gamma(n) \qquad \text{A-2}$$

If n is a positive integer,

$$\Gamma(n+1) = n! \qquad \text{A-3}$$

The binomial coefficient can be extended to non-integer values as well:

$$\binom{n}{k} = \frac{n!}{k!\,(n-k)!} = \frac{\Gamma(n+1)}{\Gamma(k+1)\Gamma(n-k+1)} \qquad \text{A-4}$$

A.2 Incomplete Gamma Function

The incomplete gamma functions have the same integrand as the Gamma function, but the limits of integration are variables. There are two incomplete Gamma functions, one for each of the two limits of integration:

$$\Gamma(n,x) = \int_x^\infty z^{n-1}e^{-z}dz \qquad \text{A-5}$$

$$\gamma(n,x) = \int_0^x z^{n-1}e^{-z}dz \qquad \text{A-6}$$

These are the upper and lower incomplete Gamma functions respectively. From the definitions we have

$$\Gamma(n) = \Gamma(n,x) + \gamma(n,x) \qquad \text{A-7}$$

A.3 Beta Function

The Beta function is a function of two variables that may be written as the integral

$$B(\alpha, \beta) = \int_0^1 z^{\alpha-1}(1-z)^{\beta-1} dz \qquad \text{A-8}$$

The Beta function may be related to the Gamma function through

$$B(\alpha, \beta) = \frac{\Gamma(\alpha)\Gamma(\beta)}{\Gamma(\alpha+\beta)} \qquad \text{A-9}$$

The Beta function can also be related to the binomial coefficient

$$\binom{n}{k} = \frac{n!}{k!\,(n-k)!} = \frac{\Gamma(n+1)}{\Gamma(k+1)\Gamma(n-k+1)} \qquad \text{A-10}$$

$$= \frac{1}{(n+1)B(n-k+1,k+1)} \qquad \text{A-11}$$

A.4 Pochhammer Symbol

The Pochhammer symbol is defined as

$$(a)_n = \frac{\Gamma(a+n)}{\Gamma(a)} = a(a+1)(a+2)\dots(a+n-1) \qquad \text{A-12}$$

A.5 Hypergeometric Functions

The hypergeometric function has the series form

$$F(\alpha, \beta, \gamma; z) = 1 + \frac{\alpha\beta}{\gamma \cdot 1}z + \frac{\alpha(\alpha+1)\beta(\beta+1)}{\gamma(\gamma+1)\cdot 1 \cdot 2}z^2$$
$$+ \frac{\alpha(\alpha+1)(\alpha+2)\beta(\beta+1)(\beta+2)}{\gamma(\gamma+1)(\gamma+2)\cdot 1 \cdot 2 \cdot 3}z^3 + \cdots \qquad \text{A-13}$$

Alternatively, the hypergeometric function can be written in integral form as

$$F(\alpha, \beta, \gamma; k) = \frac{1}{B(\beta, \gamma-\beta)}\int_0^1 z^{\beta-1}(1-z)^{\gamma-\beta-1}(1-kz)^{-\alpha} dz \qquad \text{A-14}$$
$$\text{Re}\,\gamma > \text{Re}\,\beta > 0$$

The hypergeometric function can be extended to two variables as

$$F_1(\alpha, \beta, \beta', \gamma; x, y) = \sum_{m=0}^{\infty} \sum_{n=0}^{\infty} \frac{(\alpha)_{m+n}(\beta)_m(\beta')_n}{(\gamma)_{m+n} m! \, n!} x^m y^n \qquad \text{A-15}$$

This can be written as the integral

$$F_1(\lambda, \rho, \sigma, \lambda + \mu; u, v) = \frac{1}{B(\mu, \lambda)} \int_0^1 x^{\lambda-1}(1-w)^{\mu-1}(1-uw)^{-\rho}(1-vw)^{-\sigma} dx \qquad \text{A-16}$$

A.6 Binomial Sums

$$\sum_{k=0}^{n} \binom{n}{k} x^k y^{n-k} = (x+y)^n \qquad \text{A-17}$$

$$\sum_{k=0}^{n} \binom{n}{k} (k+1) = 2^{n-1}(n+2) \qquad \text{A-18}$$

$$\sum_{k=0}^{n} \binom{n}{k}^2 = \binom{2n}{n} \qquad \text{A-19}$$

Appendix B: Binary Classifier Quick Reference

Actual Value

		A	Ā
Predicted Outcome	A	TP (True Positive)	FP (False Positive)
	Ā	FN (False Negative)	TN (True Negative)

Variable	Expression
N	Number of inputs in the test set
ρ_A	Proportion of inputs in the test set that belong to class A
φ	$\dfrac{TP}{TP + FN}$
χ	$\dfrac{FP}{FP + TN}$

Metric Name	Abb.	Prob.	Definition
True Positive	TP	-	Count of the number of times the classifier is correct when the classifier determines an input belongs to the category.
False Positive	FP	-	Count of the number of times the classifier is not correct when the classifier determines an input belongs to the category.
False Negative	FN	-	Count of the number of times the classifier is not correct when the classifier determines an input does not belong to the category.
True Negative	TN	-	Count of the number of times the classifier is correct when the classifier determines an input does not belong to the category.
Actual Positive	AP	-	$AP = TP + FN$
Actual Negative	AN	-	$AN = FP + TN$
Classified Positive	CP	-	$CP = TP + FP$
Classified Negative	CN	-	$CN = FN + TN$
True Positive Rate	TPR	$P(C_A|T_A)$	$TPR = \dfrac{TP}{TP + FN}$
False Positive Rate	FPR	$P(C_A|T_{\bar{A}})$	$FPR = \dfrac{FP}{FP + TN}$
Accuracy	ACC	-	$ACC = \dfrac{TN + TP}{N}$
True Negative Rate	TNR	$P(C_{\bar{A}}|T_{\bar{A}})$	$TNR = \dfrac{TN}{FP + TN}$
False Negative Rate	FNR	$P(C_{\bar{A}}|T_A)$	$FNR = \dfrac{FN}{TP + FN}$
Positive Predictive Value	PPV	$P(T_A|C_A)$	$PPV = \dfrac{TP}{TP + FP}$
Negative Predictive Value	NPV	$P(T_{\bar{A}}|C_{\bar{A}})$	$NPV = \dfrac{TN}{TN + FN}$
False Discovery Rate	FDR	$P(T_{\bar{A}}|C_A)$	$FDR = \dfrac{FP}{FP + TP}$
Non-Discovery Rate	NDR	$P(T_A|C_{\bar{A}})$	$NDR = \dfrac{FN}{TN + FN}$
Matthews Correlation Coefficient	MCC	-	$\dfrac{(TP x TN - FP x FN)}{\sqrt{(TP + FP)(TP + FN)(TN + FP)(TN + FN)}}$
F1 Score	F1	-	$F_1 = 2\dfrac{PPV x TPR}{PPV + TPR}$

Metric	Formula	Metric	Formula
TP	$\rho\varphi N$	FP	$(1-\rho)\chi N$
FN	$\rho(1-\varphi)N$	TN	$(1-\rho)(1-\chi)N$
AP	ρN	AN	$(1-\rho)N$
CP	$[\rho\varphi + (1-\rho)\chi]N$	CN	$[\rho(1-\varphi) + (1-\rho)(1-\chi)]N$
TPR	φ	PPV	$\dfrac{\varphi}{\varphi + (\rho^{-1}-1)\chi}$
FPR	χ	NPV	$\dfrac{1-\chi}{1-\chi + \dfrac{\rho}{1-\rho}(1-\varphi)}$
ACC	$\rho\varphi + (1-\rho)(1-\chi)$	FDR	$\dfrac{\chi}{\chi + \dfrac{\rho}{1-\rho}\varphi}$
TNR	$1-\chi$	NDR	$\dfrac{1-\varphi}{1-\varphi + (\rho^{-1}-1)(1-\chi)}$
FNR	$1-\varphi$	MCC	$(\varphi-\chi)\sqrt{\dfrac{\rho(1-\rho)}{[\rho\varphi + (1-\rho)\chi][\rho(1-\varphi) + (1-\rho)(1-\chi)]}}$
F1	$\dfrac{2\varphi^2}{\varphi + \varphi^2 + (\rho^{-1}-1)\chi}$		

Variances

$$\sigma_\varphi^2 = \frac{\varphi(1-\varphi)}{\rho N} \qquad \text{B-1}$$

$$\sigma_\chi^2 = \frac{\chi(1-\chi)}{(1-\rho)N} \qquad \text{B-2}$$

Z-Score

$$z = \frac{(\varphi - \bar{\varphi})^2 + (\chi - \bar{\chi})^2}{\sqrt{(\varphi - \bar{\varphi})^2\left(\sigma_\varphi^2 + \sigma_{\bar{\varphi}}^2\right) + (\chi - \bar{\chi})^2\left(\sigma_\chi^2 + \sigma_{\bar{\chi}}^2\right)}} \qquad \text{B-3}$$

Appendix C: Trinary Classifier Quick Reference

Actual Value

		A	B	C
Predicted Outcome	A	TA (True Positive A)	FA^B (False A-B)	FA^C (False A-C)
	B	FB^A (False B-A)	TB (True Positive B)	FB^C (False B-C)
	C	FC^A (False C-A)	FC^B (False C-B)	TC (True Positive C)

Variable	Expression
N	Number of inputs in the test set
ρ_A	Proportion of inputs in the test set that belong to class A
ρ_B	Proportion of inputs in the test set that belong to class B
φ_A	$\dfrac{TA}{TA + FB^A + FC^A}$
φ_B	$\dfrac{FB^A}{TA + FB^A + FC^A}$
χ_A	$\dfrac{FA^B}{FA^B + TB + FC^B}$
χ_B	$\dfrac{TB}{FA^B + TB + FC^B}$
ψ_A	$\dfrac{FA^C}{FA^C + FB^C + TC}$
ψ_B	$\dfrac{FB^C}{FA^C + FB^C + TC}$

Variable	Expression	Base Metrics
ρ_C	Proportion of inputs in the test set that belong to class C	$1 - \rho_A - \rho_B$
φ_C	$\dfrac{FC^A}{TA + FB^A + FC^A}$	$1 - \varphi_A - \varphi_B$
χ_C	$\dfrac{FC^B}{FA^B + TB + FC^B}$	$1 - \chi_A - \chi_B$
ψ_C	$\dfrac{TC}{FA^C + FB^C + TC}$	$1 - \psi_A - \psi_B$

Input Set Metrics

$$A^A = TA + FB^A + FC^A \tag{C-1}$$

$$A^B = FA^B + TB + FC^B \tag{C-2}$$

$$A^C = FA^C + FB^C + TC \tag{C-3}$$

$$A^A = \rho_A N \tag{C-4}$$

$$A^B = \rho_B N \tag{C-5}$$

$$A^A = (1 - \rho_A - \rho_B)N \tag{C-6}$$

Classified Set Metrics

$$C^A = TA + FA^B + FA^C \tag{C-7}$$

$$C^B = FB^A + TB + FB^C \tag{C-8}$$

$$C^C = FC^A + FC^B + TC \tag{C-9}$$

$$C^A = (\rho_A \varphi_A + \rho_B \chi_A + \rho_C \psi_A)N \tag{C-10}$$

$$C^B = (\rho_A \varphi_B + \rho_B \chi_B + \rho_C \psi_B)N \tag{C-11}$$

$$C^A = (\rho_A \varphi_C + \rho_B \chi_C + \rho_C \psi_C)N \tag{C-12}$$

Confusion Matrix Metrics

$$TA = \varphi_A \rho_A N \qquad\qquad \text{C-13}$$

$$FB^A = \varphi_B \rho_A N \qquad\qquad \text{C-14}$$

$$FC^A = \varphi_C \rho_A N \qquad\qquad \text{C-15}$$

$$FA^B = \chi_A \rho_B N \qquad\qquad \text{C-16}$$

$$TB = \chi_B \rho_B N \qquad\qquad \text{C-17}$$

$$FC^B = \chi_C \rho_B N \qquad\qquad \text{C-18}$$

$$FA^C = \psi_A \rho_C N \qquad\qquad \text{C-19}$$

$$FB^C = \psi_B \rho_C N \qquad\qquad \text{C-20}$$

$$TC = \psi_C \rho_C N \qquad\qquad \text{C-21}$$

Probability Metrics

$P(T|C)$	T_A	T_B	T_C	$T_A \vee T_B$	$T_A \vee T_C$	$T_B \vee T_C$
C_A	$\dfrac{TA}{TA + FA^B + FA^C}$	$\dfrac{FA^B}{TA + FA^B + FA^C}$	$\dfrac{FA^C}{TA + FA^B + FA^C}$	$\dfrac{TA + FA^B}{TA + FA^B + FA^C}$	$\dfrac{TA + FA^C}{TA + FA^B + FA^C}$	$\dfrac{FA^B + FA^C}{TA + FA^B + FA^C}$
C_B	$\dfrac{FB^A}{FB^A + TB + FB^C}$	$\dfrac{TB}{FB^A + TB + FB^C}$	$\dfrac{FB^C}{FB^A + TB + FB^C}$	$\dfrac{FB^A + TB}{FB^A + TB + FB^C}$	$\dfrac{FB^A + FB^C}{FB^A + TB + FB^C}$	$\dfrac{TB + FB^C}{FB^A + TB + FB^C}$
C_C	$\dfrac{FC^A}{FC^A + FC^B + TC}$	$\dfrac{FC^B}{FC^A + FC^B + TC}$	$\dfrac{TC}{FC^A + FC^B + TC}$	$\dfrac{FC^A + FC^B}{FC^A + FC^B + TC}$	$\dfrac{FC^A + TC}{FC^A + FC^B + TC}$	$\dfrac{FC^B + TC}{FC^A + FC^B + TC}$

$P(T|C)$	T_A	T_B	T_C
$C_A \vee C_B$	$\dfrac{TA + FB^A}{TA + FA^B + FA^C + FB^A + TB + FB^C}$	$\dfrac{FA^B + TB}{TA + FA^B + FA^C + FB^A + TB + FB^C}$	$\dfrac{FA^C + FB^C}{TA + FA^B + FA^C + FB^A + TB + FB^C}$
$C_A \vee C_C$	$\dfrac{TA + FC^A}{TA + FA^B + FA^C + FC^A + FC^B + TC}$	$\dfrac{FA^B + FC^B}{TA + FA^B + FA^C + FC^A + FC^B + TC}$	$\dfrac{FA^C + TC}{TA + FA^B + FA^C + FC^A + FC^B + TC}$
$C_B \vee C_C$	$\dfrac{FB^A + FC^A}{FB^A + TB + FB^C + FC^A + FC^B + TC}$	$\dfrac{TB + FC^B}{FB^A + TB + FB^C + FC^A + FC^B + TC}$	$\dfrac{FB^C + TC}{FB^A + TB + FB^C + FC^A + FC^B + TC}$

$P(T|C)$	$T_A \vee T_B$	$T_A \vee T_C$	$T_B \vee T_C$
$C_A \vee C_B$	$\dfrac{TA + FB^A + FA^B + TB}{TA + FA^B + FA^C + FB^A + TB + FB^C}$	$\dfrac{TA + FB^A + FA^C + FB^C}{TA + FA^B + FA^C + FB^A + TB + FB^C}$	$\dfrac{FA^B + TB + FA^C + FB^C}{TA + FA^B + FA^C + FB^A + TB + FB^C}$
$C_A \vee C_C$	$\dfrac{TA + FC^A + FA^B + FC^B}{TA + FA^B + FA^C + FC^A + FC^B + TC}$	$\dfrac{TA + FC^A + FA^C + TC}{TA + FA^B + FA^C + FC^A + FC^B + TC}$	$\dfrac{FA^B + FC^B + FA^C + TC}{TA + FA^B + FA^C + FC^A + FC^B + TC}$
$C_B \vee C_C$	$\dfrac{FB^A + FC^A + TB + FC^B}{FB^A + TB + FB^C + FC^A + FC^B + TC}$	$\dfrac{FB^A + FC^A + FB^C + TC}{FB^A + TB + FB^C + FC^A + FC^B + TC}$	$\dfrac{TB + FC^B + FB^C + TC}{FB^A + TB + FB^C + FC^A + FC^B + TC}$

$P(C\mid T)$	C_A	C_B	C_C	$C_A \vee C_B$	$C_A \vee C_C$	$C_B \vee C_C$
T_A	$\dfrac{TA}{TA+FB^A+FC^A}$	$\dfrac{FB^A}{TA+FB^A+FC^A}$	$\dfrac{FC^A}{TA+FB^A+FC^A}$	$\dfrac{TA+FB^A}{TA+FB^A+FC^A}$	$\dfrac{TA+FC^A}{TA+FB^A+FC^A}$	$\dfrac{FB^A+FC^A}{TA+FB^A+FC^A}$
T_B	$\dfrac{FA^B}{FA^B+TB+FC^B}$	$\dfrac{TB}{FA^B+TB+FC^B}$	$\dfrac{FC^B}{FA^B+TB+FC^B}$	$\dfrac{FA^B+TB}{FA^B+TB+FC^B}$	$\dfrac{FA^B+FC^B}{FA^B+TB+FC^B}$	$\dfrac{TB+FC^B}{FA^B+TB+FC^B}$
T_C	$\dfrac{FA^C}{FA^C+FB^C+TC}$	$\dfrac{FB^C}{FA^C+FB^C+TC}$	$\dfrac{TC}{FA^C+FB^C+TC}$	$\dfrac{FA^C+FB^C}{FA^C+FB^C+TC}$	$\dfrac{FA^C+TC}{FA^C+FB^C+TC}$	$\dfrac{FB^C+TC}{FA^C+FB^C+TC}$

$P(C\mid T)$	C_A	C_B	C_C
$T_A \vee T_B$	$\dfrac{TA+FA^B}{TA+FB^A+FC^A+FA^B+TB+FC^B}$	$\dfrac{FB^A+TB}{TA+FB^A+FC^A+FA^B+TB+FC^B}$	$\dfrac{FC^A+FC^B}{TA+FB^A+FC^A+FA^B+TB+FC^B}$
$T_A \vee T_C$	$\dfrac{TA+FA^C}{TA+FB^A+FC^A+FA^C+FB^C+TC}$	$\dfrac{FB^A+FB^C}{TA+FB^A+FC^A+FA^C+FB^C+TC}$	$\dfrac{FC^A+TC}{TA+FB^A+FC^A+FA^C+FB^C+TC}$
$T_B \vee T_C$	$\dfrac{FA^B+FA^C}{FA^B+TB+FC^B+FA^C+FB^C+TC}$	$\dfrac{TB+FB^C}{FA^B+TB+FC^B+FA^C+FB^C+TC}$	$\dfrac{FC^B+TC}{FA^B+TB+FC^B+FA^C+FB^C+TC}$

$P(C\mid T)$	$C_A \vee C_B$	$C_A \vee C_C$	$C_B \vee C_C$
$T_A \vee T_B$	$\dfrac{TA+FA^B+FB^A+TB}{TA+FB^A+FC^A+FA^B+TB+FC^B}$	$\dfrac{TA+FA^B+FC^A+FC^B}{TA+FB^A+FC^A+FA^B+TB+FC^B}$	$\dfrac{FB^A+TB+FC^A+FC^B}{TA+FB^A+FC^A+FA^B+TB+FC^B}$
$T_A \vee T_C$	$\dfrac{TA+FA^C+FB^A+FB^C}{TA+FB^A+FC^A+FA^C+FB^C+TC}$	$\dfrac{TA+FA^C+FC^A+TC}{TA+FB^A+FC^A+FA^C+FB^C+TC}$	$\dfrac{FB^A+FB^C+FC^A+TC}{TA+FB^A+FC^A+FA^C+FB^C+TC}$
$T_B \vee T_C$	$\dfrac{FA^B+FA^C+TB+FB^C}{FA^B+TB+FC^B+FA^C+FB^C+TC}$	$\dfrac{FA^B+FA^C+FC^B+TC}{FA^B+TB+FC^B+FA^C+FB^C+TC}$	$\dfrac{TB+FB^C+FC^B+TC}{FA^B+TB+FC^B+FA^C+FB^C+TC}$

$P(T\mid C)$	T_A	T_B	T_C	$T_A \vee T_B$	$T_A \vee T_C$	$T_B \vee T_C$
C_A	$\dfrac{\varphi_A\rho_A}{\varphi_A\rho_A+\chi_A\rho_B+\psi_A\rho_C}$	$\dfrac{\chi_A\rho_B}{\varphi_A\rho_A+\chi_A\rho_B+\psi_A\rho_C}$	$\dfrac{\psi_A\rho_C}{\varphi_A\rho_A+\chi_A\rho_B+\psi_A\rho_C}$	$\dfrac{\varphi_A\rho_A+\chi_A\rho_B}{\varphi_A\rho_A+\chi_A\rho_B+\psi_A\rho_C}$	$\dfrac{\varphi_A\rho_A+\psi_A\rho_C}{\varphi_A\rho_A+\chi_A\rho_B+\psi_A\rho_C}$	$\dfrac{\chi_A\rho_B+\psi_A\rho_C}{\varphi_A\rho_A+\chi_A\rho_B+\psi_A\rho_C}$
C_B	$\dfrac{\varphi_B\rho_A}{\varphi_B\rho_A+\chi_B\rho_B+\psi_B\rho_C}$	$\dfrac{\chi_B\rho_B}{\varphi_B\rho_A+\chi_B\rho_B+\psi_B\rho_C}$	$\dfrac{\psi_B\rho_C}{\varphi_B\rho_A+\chi_B\rho_B+\psi_B\rho_C}$	$\dfrac{\varphi_B\rho_A+\chi_B\rho_B}{\varphi_B\rho_A+\chi_B\rho_B+\psi_B\rho_C}$	$\dfrac{\varphi_B\rho_A+\psi_B\rho_C}{\varphi_B\rho_A+\chi_B\rho_B+\psi_B\rho_C}$	$\dfrac{\chi_B\rho_B+\psi_B\rho_C}{\varphi_B\rho_A+\chi_B\rho_B+\psi_B\rho_C}$
C_C	$\dfrac{\varphi_C\rho_A}{\varphi_C\rho_A+\chi_C\rho_B+\psi_C\rho_C}$	$\dfrac{\chi_C\rho_B}{\varphi_C\rho_A+\chi_C\rho_B+\psi_C\rho_C}$	$\dfrac{\psi_C\rho_C}{\varphi_C\rho_A+\chi_C\rho_B+\psi_C\rho_C}$	$\dfrac{\varphi_C\rho_A+\chi_C\rho_B}{\varphi_C\rho_A+\chi_C\rho_B+\psi_C\rho_C}$	$\dfrac{\varphi_C\rho_A+\psi_C\rho_C}{\varphi_C\rho_A+\chi_C\rho_B+\psi_C\rho_C}$	$\dfrac{\chi_C\rho_B+\psi_C\rho_C}{\varphi_C\rho_A+\chi_C\rho_B+\psi_C\rho_C}$

$P(T\mid C)$	T_A	T_B	T_C
$C_A \vee C_B$	$\dfrac{\varphi_A\rho_A+\varphi_B\rho_A}{\varphi_A\rho_A+\chi_A\rho_B+\psi_A\rho_C+\varphi_B\rho_A+\chi_B\rho_B+\psi_B\rho_C}$	$\dfrac{\chi_A\rho_B+\chi_B\rho_B}{\varphi_A\rho_A+\chi_A\rho_B+\psi_A\rho_C+\varphi_B\rho_A+\chi_B\rho_B+\psi_B\rho_C}$	$\dfrac{\psi_A\rho_C+\psi_B\rho_C}{\varphi_A\rho_A+\chi_A\rho_B+\psi_A\rho_C+\varphi_B\rho_A+\chi_B\rho_B+\psi_B\rho_C}$
$C_A \vee C_C$	$\dfrac{\varphi_A\rho_A+\varphi_C\rho_A}{\varphi_A\rho_A+\chi_A\rho_B+\psi_A\rho_C+\varphi_C\rho_A+\chi_C\rho_B+\psi_C\rho_C}$	$\dfrac{\chi_A\rho_B+\chi_C\rho_B}{\varphi_A\rho_A+\chi_A\rho_B+\psi_A\rho_C+\varphi_C\rho_A+\chi_C\rho_B+\psi_C\rho_C}$	$\dfrac{\psi_A\rho_C+\psi_C\rho_C}{\varphi_A\rho_A+\chi_A\rho_B+\psi_A\rho_C+\varphi_C\rho_A+\chi_C\rho_B+\psi_C\rho_C}$
$C_B \vee C_C$	$\dfrac{\varphi_B\rho_A+\varphi_C\rho_A}{\varphi_B\rho_A+\chi_B\rho_B+\psi_B\rho_C+\varphi_C\rho_A+\chi_C\rho_B+\psi_C\rho_C}$	$\dfrac{\chi_B\rho_B+\chi_C\rho_B}{\varphi_B\rho_A+\chi_B\rho_B+\psi_B\rho_C+\varphi_C\rho_A+\chi_C\rho_B+\psi_C\rho_C}$	$\dfrac{\psi_B\rho_C+\psi_C\rho_C}{\varphi_B\rho_A+\chi_B\rho_B+\psi_B\rho_C+\varphi_C\rho_A+\chi_C\rho_B+\psi_C\rho_C}$

$P(T\|C)$	$T_A \vee T_B$	$T_A \vee T_C$	$T_B \vee T_C$
$C_A \vee C_B$	$\dfrac{\varphi_A\rho_A + \varphi_B\rho_A + \chi_A\rho_B + \chi_B\rho_B}{\varphi_A\rho_A + \chi_A\rho_B + \psi_A\rho_C + \varphi_B\rho_A + \chi_B\rho_B + \psi_B\rho_C}$	$\dfrac{\varphi_A\rho_A + \varphi_B\rho_A + \psi_A\rho_C + \psi_B\rho_C}{\varphi_A\rho_A + \chi_A\rho_B + \psi_A\rho_C + \varphi_B\rho_A + \chi_B\rho_B + \psi_B\rho_C}$	$\dfrac{\chi_A\rho_B + \chi_B\rho_B + \psi_A\rho_C + \psi_B\rho_C}{\varphi_A\rho_A + \chi_A\rho_B + \psi_A\rho_C + \varphi_B\rho_A + \chi_B\rho_B + \psi_B\rho_C}$
$C_A \vee C_C$	$\dfrac{\varphi_A\rho_A + \varphi_C\rho_A + \chi_A\rho_B + \chi_C\rho_B}{\varphi_A\rho_A + \chi_A\rho_B + \psi_A\rho_C + \varphi_C\rho_A + \chi_C\rho_B + \psi_C\rho_C}$	$\dfrac{\varphi_A\rho_A + \varphi_C\rho_A + \psi_A\rho_C + \psi_C\rho_C}{\varphi_A\rho_A + \chi_A\rho_B + \psi_A\rho_C + \varphi_C\rho_A + \chi_C\rho_B + \psi_C\rho_C}$	$\dfrac{\chi_A\rho_B + \chi_C\rho_B + \psi_A\rho_C + \psi_C\rho_C}{\varphi_A\rho_A + \chi_A\rho_B + \psi_A\rho_C + \varphi_C\rho_A + \chi_C\rho_B + \psi_C\rho_C}$
$C_B \vee C_C$	$\dfrac{\varphi_B\rho_A + \varphi_C\rho_A + \chi_B\rho_B + \chi_C\rho_B}{\varphi_B\rho_A + \chi_B\rho_B + \psi_B\rho_C + \varphi_C\rho_A + \chi_C\rho_B + \psi_C\rho_C}$	$\dfrac{\varphi_B\rho_A + \varphi_C\rho_A + \psi_B\rho_C + \psi_C\rho_C}{\varphi_B\rho_A + \chi_B\rho_B + \psi_B\rho_C + \varphi_C\rho_A + \chi_C\rho_B + \psi_C\rho_C}$	$\dfrac{\chi_B\rho_B + \chi_C\rho_B + \psi_B\rho_C + \psi_C\rho_C}{\varphi_B\rho_A + \chi_B\rho_B + \psi_B\rho_C + \varphi_C\rho_A + \chi_C\rho_B + \psi_C\rho_C}$

$P(C\|T)$	C_A	C_B	C_C	$C_A \vee C_B$	$C_A \vee C_C$	$C_B \vee C_C$
T_A	φ_A	φ_B	φ_C	$\varphi_A + \varphi_B$	$\varphi_A + \varphi_C$	$\varphi_B + \varphi_C$
T_B	χ_A	χ_B	χ_C	$\chi_A + \chi_B$	$\chi_A + \chi_C$	$\chi_B + \chi_C$
T_C	ψ_A	ψ_B	ψ_C	$\psi_A + \psi_B$	$\psi_A + \psi_C$	$\psi_B + \psi_C$

$P(C\|T)$	C_A	C_B	C_C
$T_A \vee T_B$	$\dfrac{\varphi_A\rho_A + \chi_A\rho_B}{\rho_A + \rho_B}$	$\dfrac{\varphi_B\rho_A + \chi_B\rho_B}{\rho_A + \rho_B}$	$\dfrac{\varphi_C\rho_A + \chi_C\rho_B}{\rho_A + \rho_B}$
$T_A \vee T_C$	$\dfrac{\varphi_A\rho_A + \psi_A\rho_C}{\rho_A + \rho_C}$	$\dfrac{\varphi_B\rho_A + \psi_B\rho_C}{\rho_A + \rho_C}$	$\dfrac{\varphi_C\rho_A + \psi_C\rho_C}{\rho_A + \rho_C}$
$T_B \vee T_C$	$\dfrac{\chi_A\rho_B + \psi_A\rho_C}{\rho_B + \rho_C}$	$\dfrac{\chi_B\rho_B + \psi_B\rho_C}{\rho_B + \rho_C}$	$\dfrac{\chi_C\rho_B + \psi_C\rho_C}{\rho_B + \rho_C}$

$P(C\|T)$	$C_A \vee C_B$	$C_A \vee C_C$	$C_B \vee C_C$
$T_A \vee T_B$	$\dfrac{(\varphi_A + \varphi_B)\rho_A + (\chi_A + \chi_B)\rho_B}{\rho_A + \rho_B}$	$\dfrac{(\varphi_A + \varphi_C)\rho_A + (\chi_A + \chi_C)\rho_B}{\rho_A + \rho_B}$	$\dfrac{(\varphi_B + \varphi_C)\rho_A + (\chi_B + \chi_C)\rho_B}{\rho_A + \rho_B}$
$T_A \vee T_C$	$\dfrac{(\varphi_A + \varphi_B)\rho_A + (\psi_A + \psi_B)\rho_C}{\rho_A + \rho_C}$	$\dfrac{(\varphi_A + \varphi_C)\rho_A + (\psi_A + \psi_C)\rho_C}{\rho_A + \rho_C}$	$\dfrac{(\varphi_B + \varphi_C)\rho_A + (\psi_B + \psi_C)\rho_C}{\rho_A + \rho_C}$
$T_B \vee T_C$	$\dfrac{(\chi_A + \chi_B)\rho_B + (\psi_A + \psi_B)\rho_C}{\rho_B + \rho_C}$	$\dfrac{(\chi_A + \chi_C)\rho_B + (\psi_A + \psi_C)\rho_C}{\rho_B + \rho_C}$	$\dfrac{(\chi_B + \chi_C)\rho_B + (\psi_B + \psi_C)\rho_C}{\rho_B + \rho_C}$

Accuracy

$$ACC = \frac{TA + TB + TC}{N} \qquad \text{C-22}$$

$$ACC = \varphi_A\rho_A + \chi_B\rho_B + \psi_C\rho_C \qquad \text{C-23}$$

Determent

$$D = [\varphi_A(\chi_B - \psi_B) - \chi_A(\varphi_B - \psi_B) + \psi_A(\varphi_B - \chi_B)]\rho_A\rho_B\rho_C \qquad \text{C-24}$$

Positive Performance Matrix

	A	B	C
Correct	φ_A	χ_B	ψ_C
Incorrect	$1 - \varphi_A$	$1 - \chi_B$	$1 - \psi_C$

Decomposition to Binary Classifiers

$A \qquad \begin{pmatrix} TA & FA^B + FA^C \\ FB^A + FC^A & TB + FB^C + FC^B + TC \end{pmatrix}$ C-25

$B \qquad \begin{pmatrix} TB & FB^A + FB^C \\ FA^B + FC^B & TA + FA^C + FC^A + TC \end{pmatrix}$ C-26

$C \qquad \begin{pmatrix} TC & FC^A + FC^B \\ FA^C + FB^C & TA + FA^B + FB^A + TB \end{pmatrix}$ C-27

$A \qquad \begin{pmatrix} \varphi_A & \dfrac{\rho_B\chi_A + \rho_C\psi_A}{\rho_B + \rho_C} \\[2ex] 1 - \varphi_A & \dfrac{(1 - \chi_A)\rho_B + (1 - \psi_A)\rho_C}{\rho_B + \rho_C} \end{pmatrix}$ C-28

$B \qquad \begin{pmatrix} \chi_B & \dfrac{\rho_A\varphi_B + \rho_C\psi_B}{\rho_A + \rho_C} \\[2ex] 1 - \chi_B & \dfrac{(1 - \varphi_B)\rho_A + (1 - \psi_B)\rho_C}{\rho_A + \rho_C} \end{pmatrix}$ C-29

$C \qquad \begin{pmatrix} \psi_C & \dfrac{\rho_A\varphi_C + \rho_B\chi_C}{\rho_A + \rho_B} \\[2ex] 1 - \psi_C & \dfrac{(1 - \varphi_C)\rho_A + (1 - \chi_C)\rho_B}{\rho_A + \rho_B} \end{pmatrix}$ C-30

Appendix D: Multiclass Quick Reference

Confusion Matrix

$$\mathfrak{C} = \begin{pmatrix} \mathfrak{C}_{11} & \cdots & \mathfrak{C}_{1k} \\ \vdots & \ddots & \vdots \\ \mathfrak{C}_{k1} & \cdots & \mathfrak{C}_{kk} \end{pmatrix}$$

D-1

Base Matrix

$$\mathcal{B}_{ij} = \frac{\mathfrak{C}_{ij}}{\rho_j N}$$

D-2

$$\mathfrak{C}_{ij} = \mathcal{B}_{ij} \rho_j N$$

D-3

Input Set Metrics

$$N_j = \left[\mathfrak{C}_{ij}\right]_i = \rho_j N$$

D-4

Classified Set Metrics

$$C^i = \left[\mathfrak{C}_{ij}\right]_j$$

D-5

$$C^i = \left[\rho_j \mathcal{B}_{ij}\right]_j N$$

D-6

Confusion Matrix Metrics

$$\mathfrak{C}_{ij} = \mathcal{B}_{ij} \rho_j N$$

D-7

$$\mathcal{B}_{ij} = \frac{\mathfrak{C}_{ij}}{\rho_j N}$$

D-8

Probability Metrics

$$P(T_k|C_l) = \frac{\mathfrak{C}_{lk}}{[\mathfrak{C}_{li}]_i}$$

D-9

$$P(C_k|T_l) = \frac{\mathfrak{C}_{kl}}{[\mathfrak{C}_{ik}]_i}$$

D-10

$$P(T_k|C_l) = \frac{\mathcal{B}_{lk}\rho_k}{[\mathcal{B}_{li}\rho_i]_i}$$

D-11

$$P(C_k|T_l) = \mathcal{B}_{kl}$$

D-12

$$P\left(\bigvee_{i=1}^{n} T_{k_i} \mid \bigvee_{i=1}^{m} C_{l_i}\right) = \frac{\sum_{k_i=l_i} \mathfrak{C}_{l_i k_i}}{\sum_{i=1}^{m} [\mathfrak{C}_{l_i j}]_j} \tag{D-13}$$

$$P\left(\bigvee_{i=1}^{m} C_{l_i} \mid \bigvee_{i=1}^{n} T_{k_i}\right) = \frac{\sum_{k_i=l_i} \mathfrak{C}_{l_i k_i}}{\sum_{i=1}^{n} [\mathfrak{C}_{j k_i}]_j} \tag{D-14}$$

$$P\left(\bigvee_{i=1}^{n} T_{k_i} \mid \bigvee_{i=1}^{m} C_{l_i}\right) = \frac{\sum_{k_i=l_i} \mathfrak{B}_{l_i k_i} \rho_{k_i}}{\sum_{i=1}^{m} [\mathfrak{B}_{l_i j} \rho_j]_j} \tag{D-15}$$

$$P\left(\bigvee_{i=1}^{m} C_{l_i} \mid \bigvee_{i=1}^{n} T_{k_i}\right) = \frac{\sum_{k_i=l_i} \mathfrak{B}_{l_i k_i} \rho_{k_i}}{\sum_{i=1}^{n} \rho_{k_i}} \tag{D-16}$$

Accuracy

$$ACC = \frac{[\mathfrak{C}_{ii}]_i}{N} \tag{D-17}$$

$$ACC = [\mathfrak{B}_{ii} \rho_i]_i \tag{D-18}$$

Determinant

$$D = \frac{|\mathfrak{C}|}{N^k} \tag{D-19}$$

Positive Performance Matrix

$$PPM = \begin{pmatrix} \mathfrak{B}_{11} & \cdots & \mathfrak{B}_{kk} \\ 1 - \mathfrak{B}_{11} & \cdots & 1 - \mathfrak{B}_{kk} \end{pmatrix} \tag{D-20}$$

Variance

$$\sigma^2_{\mathfrak{B}_{ii}} = \frac{\mathfrak{B}_{ii}(1 - \mathfrak{B}_{ii})}{\rho_i N} \tag{D-21}$$

Overall Accuracy

$$ACC = [\mathfrak{B}_{ii} \rho_i]_i \tag{D-22}$$

Appendix E: Tables of Functions

Regularized Gamma Function

The regularized gamma function is

$$P(n, x) = \frac{1}{\Gamma(n)} \gamma(n, x) = \frac{1}{\Gamma(n)} \int_0^x t^{n-1} e^{-t} dt \qquad \text{E-1}$$

where $\gamma(n, x)$ is the lower incomplete gamma function

$$\gamma(n, x) = \int_0^x t^{n-1} e^{-t} dt \qquad \text{E-2}$$

The regularized gamma function is tied to the Hotelling statistic for treating an n-dimensional multiclass classifier as n individual binary classifiers. This statistic is used to compare two different such classifiers to determine if they are statistically different. The statistic is

$$H = \sum_{i=1}^{k} \frac{\left(\mathcal{B}_{ii} - \bar{\mathcal{B}}_{ii}\right)^2}{\sigma_{\mathcal{B}_{ii}}^2 + \sigma_{\bar{\mathcal{B}}_{ii}}^2} \qquad \text{E-3}$$

where $\mathcal{B}_{ii}$ and $\bar{\mathcal{B}}_{ii}$ are the performance metrics for the binary classifiers, and $\sigma_{\mathcal{B}_{ii}}^2$ and $\sigma_{\bar{\mathcal{B}}_{ii}}^2$ are the respective variances.

The tables that follow present the critical values for this statistic in relation to the confidence level. Given a number of dimensions and a desired confidence, we can look up the minimum value for the Hotelling statistic. In a particular experiment, if the value of the Hotelling statistic is less than the critical value, then the two classifiers are not distinct at the given confidence level.

Absolute Z-Score Sum

The Hotelling statistic examines the sum of the square of the z-score for a set of binary classifiers. Alternatively, we can examine the sum of the absolute value of the z-scores. This statistic is given by

$$K = \sum_{i=1}^{k} \frac{\left|\mathcal{B}_{ii} - \bar{\mathcal{B}}_{ii}\right|}{\sqrt{\sigma_{\mathcal{B}_{ii}}^2 + \sigma_{\bar{\mathcal{B}}_{ii}}^2}} \qquad \text{E-4}$$

Confidence (1%-25%) and Number of Terms for the Hotelling Statistic

	0.01	0.025	0.05	0.075	0.1	0.15	0.2	0.25
1	0.000	0.001	0.004	0.009	0.016	0.036	0.064	0.102
2	0.020	0.051	0.103	0.156	0.211	0.325	0.446	0.575
3	0.115	0.216	0.352	0.472	0.584	0.798	1.01	1.21
4	0.297	0.484	0.711	0.897	1.06	1.37	1.65	1.92
5	0.554	0.831	1.15	1.39	1.61	1.99	2.34	2.67
6	0.872	1.24	1.64	1.94	2.20	2.66	3.07	3.45
7	1.24	1.69	2.17	2.53	2.83	3.36	3.82	4.25
8	1.65	2.18	2.73	3.14	3.49	4.08	4.59	5.07
9	2.09	2.70	3.33	3.78	4.17	4.82	5.38	5.90
10	2.56	3.25	3.94	4.45	4.87	5.57	6.18	6.74
11	3.05	3.82	4.57	5.12	5.58	6.34	6.99	7.58
12	3.57	4.40	5.23	5.82	6.30	7.11	7.81	8.44
13	4.11	5.01	5.89	6.52	7.04	7.90	8.63	9.30
14	4.66	5.63	6.57	7.24	7.79	8.70	9.47	10.2
15	5.23	6.26	7.26	7.97	8.55	9.50	10.3	11.0
16	5.81	6.91	7.96	8.71	9.31	10.3	11.2	11.9
17	6.41	7.56	8.67	9.45	10.1	11.1	12.0	12.8
18	7.01	8.23	9.39	10.2	10.9	11.9	12.9	13.7
19	7.63	8.91	10.1	11.0	11.7	12.8	13.7	14.6
20	8.26	9.59	10.9	11.7	12.4	13.6	14.6	15.5
25	11.5	13.1	14.6	15.6	16.5	17.8	18.9	19.9
50	29.7	32.4	34.8	36.4	37.7	39.8	41.4	42.9
75	49.5	52.9	56.1	58.1	59.8	62.4	64.5	66.4
100	70.1	74.2	77.9	80.4	82.4	85.4	87.9	90.1

Table 56: Table of the regularized gamma function. The number of classifiers are designated in the first column, while the column headers provide the confidence.

Confidence (30%-65%) and Number of Terms for the Hotelling Statistic

	0.3	0.35	0.4	0.45	0.5	0.55	0.6	0.65
1	0.148	0.206	0.275	0.357	0.455	0.571	0.708	0.873
2	0.713	0.862	1.02	1.20	1.39	1.60	1.83	2.10
3	1.42	1.64	1.87	2.11	2.37	2.64	2.95	3.28
4	2.19	2.47	2.75	3.05	3.36	3.69	4.04	4.44
5	3.00	3.33	3.66	4.00	4.35	4.73	5.13	5.57
6	3.83	4.20	4.57	4.95	5.35	5.77	6.21	6.69
7	4.67	5.08	5.49	5.91	6.35	6.80	7.28	7.81
8	5.53	5.98	6.42	6.88	7.34	7.83	8.35	8.91
9	6.39	6.88	7.36	7.84	8.34	8.86	9.41	10.0
10	7.27	7.78	8.30	8.81	9.34	9.89	10.5	11.1
11	8.15	8.70	9.24	9.78	10.3	10.9	11.5	12.2
12	9.03	9.61	10.2	10.8	11.3	11.9	12.6	13.3
13	9.93	10.5	11.1	11.7	12.3	13.0	13.6	14.3
14	10.8	11.5	12.1	12.7	13.3	14.0	14.7	15.4
15	11.7	12.4	13.0	13.7	14.3	15.0	15.7	16.5
16	12.6	13.3	14.0	14.7	15.3	16.0	16.8	17.6
17	13.5	14.2	14.9	15.6	16.3	17.1	17.8	18.6
18	14.4	15.2	15.9	16.6	17.3	18.1	18.9	19.7
19	15.4	16.1	16.9	17.6	18.3	19.1	19.9	20.8
20	16.3	17.0	17.8	18.6	19.3	20.1	21.0	21.8
25	20.9	21.8	22.6	23.5	24.3	25.2	26.1	27.1
50	44.3	45.6	46.9	48.1	49.3	50.6	51.9	53.3
75	68.1	69.7	71.3	72.8	74.3	75.9	77.5	79.1
100	92.1	94.0	95.8	97.6	99.3	101	103	105

Table 57: Table of the regularized gamma function. The number of classifiers are designated in the first column, while the column headers provide the confidence.

Confidence (70%-97.5%) and Number of Terms for the Hotelling Statistic

	0.7	0.75	0.8	0.85	0.9	0.925	0.95	0.975
1	1.07	1.32	1.64	2.07	2.71	3.17	3.84	5.02
2	2.41	2.77	3.22	3.79	4.61	5.18	5.99	7.38
3	3.66	4.11	4.64	5.32	6.25	6.90	7.81	9.35
4	4.88	5.39	5.99	6.74	7.78	8.50	9.49	11.1
5	6.06	6.63	7.29	8.12	9.24	10.0	11.1	12.8
6	7.23	7.84	8.56	9.45	10.6	11.5	12.6	14.4
7	8.38	9.04	9.80	10.7	12.0	12.9	14.1	16.0
8	9.52	10.2	11.0	12.0	13.4	14.3	15.5	17.5
9	10.7	11.4	12.2	13.3	14.7	15.6	16.9	19.0
10	11.8	12.5	13.4	14.5	16.0	17.0	18.3	20.5
11	12.9	13.7	14.6	15.8	17.3	18.3	19.7	21.9
12	14.0	14.8	15.8	17.0	18.5	19.6	21.0	23.3
13	15.1	16.0	17.0	18.2	19.8	20.9	22.4	24.7
14	16.2	17.1	18.2	19.4	21.1	22.2	23.7	26.1
15	17.3	18.2	19.3	20.6	22.3	23.5	25.0	27.5
16	18.4	19.4	20.5	21.8	23.5	24.7	26.3	28.8
17	19.5	20.5	21.6	23.0	24.8	26.0	27.6	30.2
18	20.6	21.6	22.8	24.2	26.0	27.2	28.9	31.5
19	21.7	22.7	23.9	25.3	27.2	28.5	30.1	32.9
20	22.8	23.8	25.0	26.5	28.4	29.7	31.4	34.2
25	28.2	29.3	30.7	32.3	34.4	35.8	37.7	40.6
50	54.7	56.3	58.2	60.3	63.2	65.0	67.5	71.4
75	80.9	82.9	85.1	87.7	91.1	93.3	96.2	101
100	107	109	112	115	118	121	124	130

Table 58: Table of the regularized gamma function. The number of classifiers are designated in the first column, while the column headers provide the confidence.

Confidence (98%-99.75%) and Number of Terms for the Hotelling Statistic

	0.98	0.9825	0.985	0.9875	0.99	0.9925	0.995	0.9975
1	5.41	5.65	5.92	6.24	6.63	7.15	7.88	9.14
2	7.82	8.09	8.40	8.76	9.21	9.79	10.6	12.0
3	9.84	10.1	10.5	10.9	11.3	12.0	12.8	14.3
4	11.7	12.0	12.3	12.8	13.3	13.9	14.9	16.4
5	13.4	13.7	14.1	14.5	15.1	15.8	16.7	18.4
6	15.0	15.4	15.8	16.2	16.8	17.5	18.5	20.2
7	16.6	17.0	17.4	17.9	18.5	19.2	20.3	22.0
8	18.2	18.5	19.0	19.5	20.1	20.9	22.0	23.8
9	19.7	20.1	20.5	21.0	21.7	22.5	23.6	25.5
10	21.2	21.6	22.0	22.6	23.2	24.0	25.2	27.1
11	22.6	23.0	23.5	24.1	24.7	25.6	26.8	28.7
12	24.1	24.5	25.0	25.5	26.2	27.1	28.3	30.3
13	25.5	25.9	26.4	27.0	27.7	28.6	29.8	31.9
14	26.9	27.3	27.8	28.4	29.1	30.1	31.3	33.4
15	28.3	28.7	29.2	29.8	30.6	31.5	32.8	34.9
16	29.6	30.1	30.6	31.3	32.0	33.0	34.3	36.5
17	31.0	31.5	32.0	32.6	33.4	34.4	35.7	37.9
18	32.3	32.8	33.4	34.0	34.8	35.8	37.2	39.4
19	33.7	34.2	34.7	35.4	36.2	37.2	38.6	40.9
20	35.0	35.5	36.1	36.8	37.6	38.6	40.0	42.3
25	41.6	42.1	42.7	43.4	44.3	45.4	46.9	49.4
50	72.6	73.3	74.1	75.0	76.2	77.6	79.5	82.7
75	102	103	104	105	106	108	110	114
100	131	132	133	134	136	138	140	144

Table 59: Table of the regularized gamma function. The number of classifiers are designated in the first column, while the column headers provide the confidence.

Confidence (99.9%-99.999%) and Number of Terms for the Hotelling Statistic

	0.999	0.9999	0.99999
1	10.8	15.1	19.5
2	13.8	18.4	23.0
3	16.3	21.1	25.9
4	18.5	23.5	28.5
5	20.5	25.7	30.9
6	22.5	27.9	33.1
7	24.3	29.9	35.3
8	26.1	31.8	37.3
9	27.9	33.7	39.3
10	29.6	35.6	41.3
11	31.3	37.4	43.2
12	32.9	39.1	45.1
13	34.5	40.9	46.9
14	36.1	42.6	48.7
15	37.7	44.3	50.5
16	39.3	45.9	52.2
17	40.8	47.6	54.0
18	42.3	49.2	55.7
19	43.8	50.8	57.4
20	45.3	52.4	59.0
25	52.6	60.1	67.2
50	86.7	96.0	105
75	119	129	139
100	149	161	172

Table 60: Table of the regularized gamma function. The number of classifiers are designated in the first column, while the column headers provide the confidence.

Confidence (1%-25%) and Number of Terms for the Absolute Statistic

	0.01	0.025	0.05	0.075	0.1	0.15	0.2	0.25
1	0.013	0.031	0.063	0.094	0.126	0.189	0.253	0.319
2	0.178	0.282	0.402	0.495	0.576	0.716	0.840	0.954
3	0.497	0.681	0.871	1.01	1.12	1.32	1.48	1.63
4	0.901	1.15	1.40	1.57	1.71	1.95	2.14	2.32
5	1.36	1.66	1.96	2.16	2.33	2.60	2.82	3.02
6	1.85	2.21	2.54	2.78	2.96	3.26	3.51	3.73
7	2.37	2.77	3.14	3.40	3.61	3.94	4.21	4.45
8	2.90	3.35	3.76	4.04	4.26	4.62	4.92	5.17
9	3.46	3.94	4.39	4.69	4.93	5.31	5.63	5.90
10	4.02	4.55	5.02	5.35	5.60	6.01	6.34	6.64
11	4.60	5.16	5.67	6.01	6.28	6.71	7.06	7.37
12	5.19	5.78	6.32	6.68	6.96	7.42	7.79	8.11
13	5.79	6.41	6.98	7.35	7.65	8.13	8.51	8.85
14	6.40	7.05	7.64	8.03	8.35	8.84	9.24	9.59
15	7.01	7.69	8.31	8.72	9.04	9.55	9.97	10.3
16	7.63	8.34	8.98	9.41	9.74	10.3	10.7	11.1
17	8.25	8.99	9.65	10.1	10.4	11.0	11.4	11.8
18	8.88	9.65	10.3	10.8	11.1	11.7	12.2	12.6
19	9.52	10.3	11.0	11.5	11.9	12.4	12.9	13.3
20	10.2	11.0	11.7	12.2	12.6	13.2	13.7	14.1
25	13.4	14.3	15.2	15.7	16.1	16.8	17.4	17.9
50	30.4	31.8	33.1	33.9	34.5	35.5	36.3	37.0
75	48.2	49.9	51.4	52.4	53.2	54.4	55.4	56.3
100	66.2	68.3	70.0	71.2	72.1	73.5	74.7	75.7

Table 61: The number of classifiers are designated in the first column, while the column headers provide the confidence.

Confidence (30%-65%) and Number of Terms for the Absolute Statistic

	0.3	0.35	0.4	0.45	0.5	0.55	0.6	0.65
1	0.385	0.454	0.524	0.598	0.674	0.755	0.842	0.935
2	1.06	1.17	1.27	1.38	1.49	1.60	1.71	1.84
3	1.76	1.90	2.03	2.16	2.29	2.42	2.56	2.71
4	2.48	2.64	2.79	2.94	3.09	3.24	3.40	3.57
5	3.20	3.38	3.55	3.72	3.89	4.06	4.24	4.42
6	3.94	4.13	4.32	4.50	4.68	4.87	5.07	5.27
7	4.67	4.88	5.08	5.28	5.48	5.69	5.90	6.12
8	5.41	5.64	5.85	6.07	6.28	6.50	6.72	6.96
9	6.16	6.40	6.63	6.85	7.08	7.31	7.55	7.79
10	6.90	7.16	7.40	7.64	7.88	8.12	8.37	8.63
11	7.65	7.92	8.17	8.43	8.68	8.93	9.19	9.46
12	8.40	8.68	8.95	9.21	9.47	9.74	10.0	10.3
13	9.16	9.45	9.73	10.0	10.3	10.5	10.8	11.1
14	9.91	10.2	10.5	10.8	11.1	11.4	11.6	12.0
15	10.7	11.0	11.3	11.6	11.9	12.2	12.5	12.8
16	11.4	11.7	12.1	12.4	12.7	13.0	13.3	13.6
17	12.2	12.5	12.8	13.2	13.5	13.8	14.1	14.4
18	12.9	13.3	13.6	13.9	14.3	14.6	14.9	15.3
19	13.7	14.1	14.4	14.7	15.1	15.4	15.7	16.1
20	14.5	14.8	15.2	15.5	15.9	16.2	16.5	16.9
25	18.3	18.7	19.1	19.5	19.8	20.2	20.6	21.0
50	37.6	38.2	38.7	39.3	39.8	40.3	40.9	41.5
75	57.0	57.7	58.4	59.1	59.7	60.4	61.1	61.8
100	76.6	77.4	78.2	78.9	79.7	80.4	81.2	82.0

Table 62: The number of classifiers are designated in the first column, while the column headers provide the confidence.

Confidence (70%-97.5%) and Number of Terms for the Absolute Statistic

	0.7	0.75	0.8	0.85	0.9	0.925	0.95	0.975
1	1.04	1.15	1.28	1.44	1.64	1.78	1.96	2.24
2	1.97	2.12	2.29	2.49	2.76	2.93	3.16	3.53
3	2.87	3.05	3.25	3.49	3.80	4.00	4.28	4.70
4	3.75	3.95	4.18	4.45	4.80	5.03	5.34	5.82
5	4.63	4.85	5.10	5.40	5.78	6.04	6.37	6.90
6	5.49	5.73	6.00	6.33	6.74	7.02	7.38	7.95
7	6.35	6.61	6.90	7.25	7.69	7.99	8.37	8.98
8	7.21	7.48	7.79	8.16	8.63	8.94	9.35	10.0
9	8.06	8.35	8.68	9.06	9.56	9.89	10.3	11.0
10	8.91	9.21	9.56	9.96	10.5	10.8	11.3	12.0
11	9.75	10.1	10.4	10.9	11.4	11.8	12.2	13.0
12	10.6	10.9	11.3	11.7	12.3	12.7	13.2	13.9
13	11.4	11.8	12.2	12.6	13.2	13.6	14.1	14.9
14	12.3	12.6	13.0	13.5	14.1	14.5	15.0	15.9
15	13.1	13.5	13.9	14.4	15.0	15.4	16.0	16.8
16	14.0	14.3	14.8	15.3	15.9	16.3	16.9	17.8
17	14.8	15.2	15.6	16.1	16.8	17.2	17.8	18.7
18	15.6	16.0	16.5	17.0	17.7	18.1	18.7	19.6
19	16.5	16.9	17.3	17.9	18.6	19.0	19.6	20.6
20	17.3	17.7	18.2	18.8	19.5	19.9	20.6	21.5
25	21.5	21.9	22.5	23.1	23.9	24.4	25.1	26.1
50	42.1	42.7	43.5	44.3	45.4	46.1	47.1	48.5
75	62.5	63.3	64.2	65.3	66.6	67.5	68.6	70.4
100	82.9	83.8	84.8	86.0	87.6	88.6	89.9	91.9

Table 63: The number of classifiers are designated in the first column, while the column headers provide the confidence.

Confidence (98%-99.75%) and Number of Terms for the Absolute Statistic

	0.98	0.9825	0.985	0.9875	0.99	0.9925	0.995	0.9975
1	2.33	2.38	2.43	2.50	2.58	2.67	2.81	3.02
2	3.64	3.71	3.78	3.87	3.97	4.10	4.27	4.56
3	4.83	4.91	5.00	5.10	5.22	5.37	5.58	5.92
4	5.97	6.05	6.15	6.26	6.40	6.57	6.80	7.18
5	7.06	7.15	7.26	7.38	7.53	7.72	7.97	8.38
6	8.12	8.22	8.34	8.47	8.63	8.83	9.10	9.55
7	9.16	9.27	9.39	9.54	9.70	9.92	10.2	10.7
8	10.2	10.3	10.4	10.6	10.8	11.0	11.3	11.8
9	11.2	11.3	11.5	11.6	11.8	12.0	12.4	12.9
10	12.2	12.3	12.5	12.6	12.8	13.1	13.4	14.0
11	13.2	13.3	13.5	13.6	13.8	14.1	14.5	15.0
12	14.2	14.3	14.5	14.6	14.8	15.1	15.5	16.1
13	15.1	15.3	15.4	15.6	15.8	16.1	16.5	17.1
14	16.1	16.3	16.4	16.6	16.8	17.1	17.5	18.1
15	17.1	17.2	17.4	17.6	17.8	18.1	18.5	19.2
16	18.0	18.2	18.4	18.6	18.8	19.1	19.5	20.2
17	19.0	19.1	19.3	19.5	19.8	20.1	20.5	21.2
18	19.9	20.1	20.3	20.5	20.7	21.0	21.5	22.2
19	20.9	21.0	21.2	21.4	21.7	22.0	22.5	23.2
20	21.8	22.0	22.2	22.4	22.7	23.0	23.4	24.2
25	26.4	26.6	26.8	27.1	27.4	27.7	28.2	29.1
50	49.0	49.2	49.5	49.8	50.2	50.7	51.4	52.5
75	70.9	71.2	71.5	71.9	72.4	73.0	73.8	75.2
100	92.5	92.8	93.2	93.7	94.2	94.9	95.9	97.4

Table 64: The number of classifiers are designated in the first column, while the column headers provide the confidence.

Confidence (99.9%-99.999%) and Number of Terms for the Absolute Statistic

	0.999	0.9999	0.99999
1	3.29	3.89	4.42
2	4.92	5.73	6.45
3	6.34	7.30	8.16
4	7.65	8.73	9.69
5	8.90	10.1	11.1
6	10.1	11.4	12.5
7	11.3	12.6	13.8
8	12.4	13.8	15.1
9	13.5	15.0	16.4
10	14.6	16.2	17.6
11	15.7	17.3	18.8
12	16.8	18.5	20.0
13	17.9	19.6	21.2
14	18.9	20.7	22.3
15	20.0	21.8	23.4
16	21.0	22.9	24.6
17	22.0	24.0	25.7
18	23.1	25.0	26.8
19	24.1	26.1	27.9
20	25.1	27.2	29.0
25	30.1	32.3	34.4
50	53.9	57.0	59.7
75	76.8	80.5	83.7
100	99.2	103	107

Table 65: The number of classifiers are designated in the first column, while the column headers provide the confidence.

Bibliography

Bayesian Classifiers

Duda, Hart, and Stork, *Pattern Classification*, Wiley, 2000

Jensen and Graven-Nielsen, *Bayesian Networks and Decision Graphs*, Springer, 2010

Korb and Nicholson, *Bayesian Artificial Intelligence, Second Edition*, Chapman & Hall, 2010

Support Vector Machines

Abe, *Support Vector Machines for Pattern Classification*, Springer, 2010

Christmann, *Support Vector Machines*, Springer, 2008

Hamel, *Knowledge Discovery with Support Vector Machines*, Wiley, 2009

Wang, *Support Vector Machines: Theory and Applications*, Springer, 2005

Neural Networks

Bishop, *Neural Networks for Pattern Recognition*, Oxford University Press, 1996

Rao, *C++ Neural Networks and Fuzzy Logic*, BPB Publications, 2003

Ripley, *Pattern Recognition and Neural Networks*, Cambridge University Press, 2008

Fuzzy Logic

Nguyen and Walker, *A first Course in Fuzzy Logic, Third Edition*, Chapman and Hall, 2005

Klir and Yuan, *Fuzzy Sets and Fuzzy Logic, Third Edition*, Prentince Hall, 1995

Hidden Markov Models

Zucchini and MacDonald, *Hidden Markov Models for Time Series, Third Edition*, Chapman and Hall, 2009

Fink, *Markov Models for Pattern Recognition*, Springer, 2007

Index